BURLEIGH DODDS SCIENCE: INSTANT INSIGHTS

NUMBER 83

Improving the welfare of gilts and sows

I0028165

burleigh dodds
SCIENCE PUBLISHING

Published by Burleigh Dodds Science Publishing Limited
82 High Street, Sawston, Cambridge CB22 3HJ, UK
www.bdspublishing.com

Burleigh Dodds Science Publishing, 1518 Walnut Street, Suite 900, Philadelphia, PA 19102-3406, USA

First published 2023 by Burleigh Dodds Science Publishing Limited
© Burleigh Dodds Science Publishing, 2023. All rights reserved.

British Library Cataloguing in Publication Data
A catalogue record for this book is available from the British Library

ISBN 978-1-80146-639-4 (Print)
ISBN 978-1-80146-640-0 (ePub)

DOI: 10.19103/9781801466400

Typeset by Deanta Global Publishing Services, Dublin, Ireland

Contents

Series list

Title	Series number
Sweetpotato	01
Fusarium in cereals	02
Vertical farming in horticulture	03
Nutraceuticals in fruit and vegetables	04
Climate change, insect pests and invasive species	05
Metabolic disorders in dairy cattle	06
Mastitis in dairy cattle	07
Heat stress in dairy cattle	08
African swine fever	09
Pesticide residues in agriculture	10
Fruit losses and waste	11
Improving crop nutrient use efficiency	12
Antibiotics in poultry production	13
Bone health in poultry	14
Feather-pecking in poultry	15
Environmental impact of livestock production	16
Pre- and probiotics in pig nutrition	17
Improving piglet welfare	18
Crop biofortification	19
Crop rotations	20
Cover crops	21
Plant growth-promoting rhizobacteria	22
Arbuscular mycorrhizal fungi	23
Nematode pests in agriculture	24
Drought-resistant crops	25
Advances in detecting and forecasting crop pests and diseases	26
Mycotoxin detection and control	27
Mite pests in agriculture	28
Supporting cereal production in sub-Saharan Africa	29
Lameness in dairy cattle	30
Infertility/reproductive disorders in dairy cattle	31
Alternatives to antibiotics in pig production	32
Integrated crop–livestock systems	33
Genetic modification of crops	34

Chapter 1

Optimizing the health of gilts and sows during pregnancy and parturition

S. Björkman, C. Oliviero and O. A. T. Peltoniemi, University of Helsinki, Finland

1 Introduction

Current knowledge of factors affecting gilts' and sows' health during and after puberty and during pregnancy, parturition and after weaning is reviewed. Megatrends such as climate change, urbanization and globalization will eventually have a large impact on how pigs are kept and how healthy they will be in the long run. The current housing and breeding of gilts and sows worldwide need to be critically reviewed in light of megatrends and climatic changes. Because these changes are not likely to disappear, they must be considered when setting goals for breeding and designing new buildings in pig production.

Resilience of pig breeds should be a breeding goal from the point of view of health and welfare (Peltoniemi et al., 2019; Knap and Doeschl-Wilson, 2021).

http://dx.doi.org/10.19103/AS.2022.0103.15

While the present breeds are geared to produce high numbers of piglets in a litter, they might not be able to cope with emerging pathogens and megatrends such as the warming climate conditions. Also, technology is applied to improve the production environment; however, the technology used should be sustainable from the point of view of the environment and animal health and welfare.

One of the associated trends seen in the pig industry is a concomitant increase in litter size and prolongation of parturition. In the 1990s, the average duration of farrowing was 2-4 h (Madec and Leon, 1992; Jackson, 1995). Since 1990, there has been a linear increase in both (1) litter size from about 10 piglets in 1990 to close to 20 piglets in 2019 and (2) duration of farrowing from 2 h to 7-8 h (a conclusion based on 20 studies on duration of farrowing; Oliviero et al., 2019). While the described tendency is subject to differences between breeds and management (i.e. farrowing crate vs. free farrowing), the overall tendency is rather convincing. It seems that the average interval between piglets has not increased very much during the last 20 years, meaning that the extended duration of farrowing appears as an outcome of intensive breeding for prolificacy in the pig (Oliviero et al., 2019). In this review, we address the physiology behind these trends, focusing on potential management solutions to alleviate issues relating to the transfer of immunity and welfare of the sow.

2 Biosecurity issues, vaccination and deworming

At the level of an operational unit, external and internal pathogens circulate and create a threat to general health, leg health, continuation of pregnancy, successful parturition and fertility. A focus on external and internal biosecurity is essential (Neumann and Hall, 2019). It is therefore critical that gilt and sow units apply a high level of internal and external biosecurity, including a proper application of quarantine procedures when new breeding animals are bought in and are about to enter the herd (Alarcon et al., 2021).

Sows and gilts require routine vaccination and deworming programmes (Neumann and Hall, 2019). Even in well-managed indoor herds with high management standards, such as batch farrowing, all-in all-out systems and high biosecurity levels, helminth infections are still very prevalent. *Ascaris suum* is the most common helminth present on commercial farms. Sows and gilts should be dewormed about 2 weeks before moving to farrowing crates or pens. Fenbendazole is a widely used anthelmintic drug effective against the different life stages of the parasite (adults, eggs and larvae). Washing sows and gilts immediately prior to transfer to the farrowing rooms is also recommended. Depending on the country, area or farm health status, different vaccination protocols are established to maintain adequate protection against the main viral and bacterial diseases in gilts and sows. Protection against leptospira, erysipelas and parvovirus should be given either pre-farrowing or at weaning, depending on the optimal timing for the particular herd. Erysipelas can cause arthritis and death losses in the herd with a marked economic impact. Leptospirosis and erysipelas can also be zoonotic diseases. Parvovirus is extremely common in most swine herds (Oravainen et al., 2005). Protection of sows during gestation against these diseases (e.g. leptospirosis, erysipelas and parvovirus infections) can be obtained by vaccination of breeding animals at 6–7 months of age and 3–4 weeks later. Boosters would follow prior to insemination or every 6 months. Combined vaccines are available on the market.

Once or twice a year, vaccinating for influenza can boost herd protection. If influenza has been a problem in piglets historically, vaccinating sows pre-farrowing should be considered. Other pre-farrowing vaccinations may include either an *Escherichia coli* vaccine or an *E. coli* and rotavirus combination vaccine to help reduce the risk of scours in neonate piglets (Neumann and Hall, 2019).

If the herd has been at risk of contracting porcine reproductive and respiratory syndrome (PRRS), atrophic rhinitis or Clostridium, a pre-emptive vaccination should be given to sows to protect them and their litters (Zimmerman et al., 2019). A porcine circovirus 2 booster vaccination should also be administered to replacement gilts prior to breeding, as they are the most susceptible to reproductive issues. Some veterinarians suggest vaccinating previously exposed herds with a porcine epidemic diarrhoea (PED) vaccine.

At the time of vaccination, sows must be healthy. If they have a fever or show any other clinical signs of illness, the vaccine should be delayed for a few days. In case of attenuated bacterial vaccines, sows should not be treated with antibiotics 3 days before or after administration of the vaccine, as the antibiotic may kill the vaccine bacteria needed to stimulate immune responses.

Health of prepubertal gilts is affected by management of developing immunity, leg and udder health (Oravainen et al., 2005). Typically, immunity triggered by a vaccine requires repeated vaccinations to achieve adequate immunity against a pathogen (Oravainen et al., 2005). In real life, making staff members of a piggery operation aware of the basic concepts of use of a vaccine and immunity created by the vaccine is an important part of everyday work. If the staff is not sufficiently motivated and educated, the risk of misuse/improper use of vaccines will rise.

3 Production diseases causing fertility problems and reduced reproductive performance

Diseases of the mammary gland and urogenital tract such as mastitis, endometritis and cystitis are often related to intensive production systems, where breeding standard, hygiene and animal welfare are suboptimal, bringing both neonates and their dams to the edge of their metabolic and immunological competence. These so-called production diseases are usually connected, being either a cause or a result of each other. The postpartum dysgalactia syndrome (PDS) is probably the most important disease complex leading to high economic losses. These economic losses are due to an increased rate of piglet mortality as well as reduced number of piglets weaned. Further, these diseases are linked to reduced development and growth of the piglets as well as reduced fertility and longevity of sows. Bacteria involved are often non-specific and opportunistic, inhabiting the lower female reproductive tract. If immunity is compromised or the microbial balance disturbed, these bacteria can cause disease. Therefore, their incidence rate can vary a lot and can be very high depending on production system.

3.1 Vulvar discharge syndrome

The vulvar discharge syndrome (VDS) can cause considerable economic losses because of the small litter size at parturition and subsequent high piglet mortality and impaired sows' fertility (Dee, 1992). The source of pathological discharge may be reproductive or urinary tract disease, especially cystitis or endometritis (Althouse et al., 2019). The incidence of these diseases is on the rise because of an increase in the intensity of the production, mostly because of increased litter size and prolonged parturition (Björkman et al., 2018b).

The organisms underlying the signs and lesions of these infections are often similar (Dee, 1992), non-specific and opportunistic. Both gram-negative (e.g. *E. coli spp., Klebsiella spp., Proteus spp.*) and gram-positive bacteria (*Streptococcus spp., Staphylococcus spp., Enterococcus spp., A. suis*)) are usually involved (Biksi et al., 2002; Grahofer et al., 2014; Bellino et al., 2013; Boers et al., 2019)). The bacteria inhabit the normal physiological microbiota on the skin and mucosa of the lower female reproductive tract (Muirhead, 1986; D'Allaire et al., 1987; Dial and MacLachlan, 1988; de Winter et al., 1995; Glock and Bilkei, 2005; Oravainen et al., 2008; Kauffold, 2008; Tummaruk et al., 2010). They can ascend to the urinary and upper reproductive tract or enter into the bloodstream causing systemic effects whenever the infectious pressure is high and/or immunity of the sow is low (Oravainen et al., 2006; Bellino et al., 2013). Sows' immunity can be challenged during parturition, stressful situations, mycotoxin exposure and/or hormonal down regulation (Althouse et al., 2019). Especially hygiene, husbandry and management around the time of parturition and breeding are most important. Risk factors for puerperal endometritis are prolonged parturition, obstetrical intervention, impaired expulsion of placenta and birth of more than two stillborn piglets (Björkman et al., 2018b; Grahofer et al., 2019). Risk factors for non-puerperal endometritis are wrong time of insemination, more than two inseminations and confinement in crates, poor hygiene and lack of enrichment material (Oravainen et al., 2006, 2008). Further, any factor that causes a reduction in the frequency of urination (decreased water intake, decreased urination, incomplete emptying of bladder, etc.) should be considered important in the development of cystitis (Drolet, 2019). Therefore, the availability of water and water flow as well as water quality in terms of bacteria and physical/chemical properties need to be frequently controlled (Drolet, 2019). Further, gilts and sows need to be in good body condition, stress-free and in good health (e.g. leg health) (Drolet, 2019).

If VDS is suspected in gilts or sows, for example, because of increased birth complications, increased incidence of postpartum diseases or decreased fertility, proper diagnostics with a systematic approach needs to be applied. The diagnostic procedure addressing herd fertility problems is comprehensive and must include a critical assessment of the tripartite contributors to herd reproductive performance: (1) farrowing personnel, (2) breeding personnel and (3) sow and semen management (Althouse et al., 2019). Diagnosticians must be aware of the multitude of factors and their inter-relatedness that might be causative of the problem, including facility type, the building environment, seasonality, nutrition, sanitation/hygiene and disease-related occurrences or interventions. A presumptive diagnosis can be based on the clinical signs. Further diagnostic methods have recently been reviewed by Grahofer et al. (2020) and Björkman and Grahofer (2020).

Both cystitis and endometritis are usually subclinical and systemic reactions are rare (Althouse et al., 2019). A body temperature of more than 39.5°C, together with clinical signs such as abnormal general behaviour (i.e. lethargy and apathy), reduced feed intake (anorexia), frequent urination and/or abnormal vaginal discharge, is indicative of VDS (Stiehler et al., 2015; Grahofer et al., 2019). The vulvar discharge of affected sows varies extensively, depending on the pathogenic microorganism, duration of infection and the stage of the oestrous cycle (Dial and MacLachlan, 1988). Physiological vaginal discharge can be observed immediately after parturition, insemination and shortly before oestrus (Muirhead, 1986; Meredith, 1991; de Winter et al., 1992; Almond et al., 2006). Increased volumes of vaginal discharge are associated with endometritis, but there is no significance between the occurrence of endometritis and the colour of the vaginal discharge (Muirhead, 1986). In animals with cystitis, the urine is often reddish-brown in colour (haematuria) with a strong door (pyuria). Presumptive diagnosis of VDS in live animals is best achieved when an abnormal amount of discharge and/or frequent micturition of blood stained and cloudy urine are observed (Drolet, 2019).

Confirmatory diagnosis of VDS can be achieved by ultrasonography (Kauffold et al., 2010) and macroscopic, microscopic and chemical and bacteriologic examination of urine and/or vaginal discharge. Commonly used methods to confirm cystitis in live animals are urinalysis and urine culture (Gmeiner, 2007). For urinalysis, it is preferred to collect spontaneous midstream urine in a transparent tube. The best time to collect urine is in the morning before feeding (Kraft et al., 2005). A red or brown colour, cloudy or turbid appearance and/or a strong ammoniac or putrid odour could indicate the presence of bacteria in the urine (Tolstrup, 2017). After the macroscopic evaluation, the urine sample can be centrifuged and the sediment microscopically evaluated at ×400 magnification (Kraft et al., 2005). Erythrocytes, leukocytes and epithelial cells are counted. A sample is considered positive when there are more than five white blood cells per visual field (Bellino et al., 2013). Furthermore, the presence of transitional epithelial cells and bacteria, and a specific gravity of the urine higher than 1.020, can be indicative of cystitis (Gmeiner, 2007; Tolstrup, 2017). After urinalysis, dipslides can be used for bacteriological evaluation. They are placed into urine for about 10 s and the bacterial growth is evaluated approximately 18-24 h later. The threshold for a urinary tract infection is 10×5 colony forming units (cfu)/mL urine (Kraft et al., 2005). Dipslides can also be submitted to the laboratory for specification of the bacteria and antimicrobial susceptibility testing. Bacteriological evaluation allows to determine the exact bacteria and antibiotic sensitivities. It should always be done if antibiotics are planned to be used in the treatment in order to decrease the risk of creating antibiotic resistance.

The most common method to diagnose endometritis in live animals is ultrasonography. In evaluation of the structure of the uterus, the parameters of fluid echogenicity, echotexture and size are measured in order to provide a comprehensive diagnosis (Kauffold and Althouse, 2007; Björkman et al., 2018b; Grahofer et al., 219; Meile et al., 2019). In a sow with acute endometritis, the uterus size as well as the echotexture are increased (Kauffold and Althouse, 2007). However, the days postpartum and the parity should be taken into account when evaluating the uterine parameters (Kauffold and Althouse, 2007; Björkman et al., 2018b). In addition, fluid echogenicity in the uterus can be used as an indicator for an exudative inflammation of the uterus (Kauffold and Althouse, 2007). For further diagnostic tests, cytology and microbiology can be used. The number and type of inflammatory cells in the endometrium per visual field in the ×400 magnification of the light microscopy gets used to classify into acute and chronic endometritis. In sows with acute endometritis, more than 20 neutrophilic granulocytes can be detected in a field (de Winter et al., 1995). In comparison, chronic endometritis is defined as the presence of more than 20 lymphocytes, plasma cells or histiocytes in a field (de Winter et al., 1995). In sows without vaginal discharge, the endometrium contains a low number of neutrophilic and eosinophilic granulocytes as well as plasma cells (Kaeoket et al., 2001; Oravainen et al., 2008). Still, it needs to be considered that only about half of the sows with endometritis show vaginal discharge. Therefore, not only in sows with clinical signs but also in subclinical sows where endometritis is suspected, ultrasonography of the uterus is indicated. Further, as with cystitis, also in case of endometritis, a bacteriological examination should be performed if the use of antibiotics is needed in order to minimize the risk of creating antibiotic resistance. The uterine swab used to collect vaginal cytology can also be used for microbiological examination and antimicrobial susceptibility testing. A speculum (with a double-guarded swab) should be used to obtain a representative sample from the uterus and to avoid contamination of the bacterial flora from the vagina (Oravainen et al., 2008; Grahofer et al., 2017).

A highly valuable tool for the accurate diagnosis of reproductive disease and disorders such as the VDS is post slaughter gross pathological examination (de Jong et al., 2014). The genital organs, together with the urinary tract and bladder, are collected from specified sows on the slaughter line at the abattoir and submitted for pathological examination. In a study conducted with 502 sows on 7 farms, uterine exudate was detected in 18% with *E. coli* being the most common finding (de Jong et al., 2014).

3.2 Postpartum dysgalactia syndrome, mastitis and severe mammary oedema

Postparturient disorders in sows represent an economically important disease complex in modern pig production worldwide (Kemper, 2020). The most important disease complex is PDS. Mastitis, endometritis and cystitis have been linked as either a cause or sign of postpartum dysgalactia syndrome (PDS) (Wendt et al., 1990; Biksi et al., 2002; Kemper, 2020). PDS is characterized by insufficient colostrum and milk production by the sow during the first days of lactation (Maes et al., 2010). As a consequence, colostrum and milk intake by piglets is reduced and therefore their mortality increased (Maes et al., 2010). Further, PDS negatively affects subsequent reproductive health of the sow (Hoy, 2006).

PDS in sows has a complex multifactorial pathophysiology, and diagnosis is therefore challenging. Whenever there is an increased incidence rate in piglet mortality or disease, or a decrease in piglet weight gain and growth rate, PDS should be suspected. This is because piglets are in their first days of life totally reliant on the sow for access to colostrum and milk which is important for their growth, development and immunity. Every alteration in both milk yield and composition has major impact on the piglets. Nevertheless, other piglet-related causes need to be excluded. Causes for piglet mortality have been reviewed by Edwards and Baxter (2015).

Besides diseases of the uterus and urinary bladder, also diseases of the mammary gland can be either a cause or result of PDS. The mammary gland, unlike the urogenital tract, is located outside the body and is therefore easily accessible for examination. The diagnosis of mastitis can be done by adspection and palpation of the mammary gland. Mastitis can be a local process, restricted to one or several glands, but can also affect all mammary complexes (Gerjets and Kemper, 2009). Infected glands typically show signs of inflammation such as severe oedema and skin congestion, therefore appearing swollen, firm and warm (Gerjets and Kemper, 2009). With many glands involved, sows develop fever and lose their appetite (Gerjets and Kemper, 2009). If needed, ultrasonography can be used in the diagnosis. Affected mammary glands provide heterogeneous and hyperechoic images (Baer and Bilkei, 2005).

As with cystitis and endometritis, usually non-specific and opportunistic organisms such as *E. coli*, *Staphylococcus spp.* and *Streptococcus spp.* are involved. The current hypothesis is that interactions between endotoxins produced by gram-negative bacteria in the gut, mammary gland and/or urogenital tract, and alterations in the immune and endocrine functions, play a central role in the development of PDS (Maes et al., 2010). This is supported by a study where periparturient sows were challenged with lipopolysaccharide (LPS) endotoxin of *E. coli* and in which sows generated symptoms similar to PDS (Nachreiner and Ginther, 1974). *E. coli* originates from the environment or can already be present in the urogenital tract of pre-partum sows or enter during or after parturition.

If there is only one or a few glands involved, local treatment such as milking the affected gland(s) during nursing or after injection of oxytocin may be sufficient. If there are multiple glands involved, and the sow shows general signs, systemic treatment with antibiotics needs to be considered. Again, in order to prevent antibiotic resistance, it is essential to take a milk sample for identification of the bacteria and antimicrobial susceptibility testing. How to take a milk sample has been described, for example, by Gerjets and Kemper (2009) and Kaiser et al (2020). Shortly, after intramuscular (IM) injection of oxytocin (20–30 I.U.), any milk present in the nipples (ductus papillaris) is removed before sampling. The teats are cleaned with, for example, 5% chlorhexidine gluconate soap solution and disinfected with, for example, 70% isopropyl alcohol. The first streams of secretion from each teat are discarded and milk samples are milked on a transport medium.

As with cystitis and endometritis, prevention needs to focus on nutrition, housing and management. Especially high levels of animal welfare and hygiene are important in order to minimize stress and infectious pressure. Risk factors for PDS are therefore the use of the unslatted floors, no washing of sows and no use of disinfectants in the farrowing and breeding rooms (Hultén et al., 2004). Considering that parturition itself decreases immunity and causes significant inflammatory changes (Kaiser et al., 2018b), all other factors affecting immunity and endocrinology need to be kept at a minimum level. Especially, stress needs to be reduced as much as possible. Elevated cortisol levels in sows with PDS were found by Kaiser et al. (2018b). Stress can be due to restricted space in farrowing crates, lack of nest-building material, high ambient temperature and abrupt change from group housing during gestation to restraint in crates a few days before farrowing (Maes et al., 2010; Papadopoulos et al., 2010).

Another disease of the mammary gland which has been lately associated with PDS is severe mammary oedema (SME, Björkman et al., 2017a, 2018a; Kaiser et al., 2020). SME can also reduce colostrum quality (Björkman et al., 2017a, 2018a). One possible explanation is that sows that will suffer from PDS may already have before and during parturition a compromised cardiovascular system and vascular homeostasis. Kaiser et al. (2020) reported increased concentrations of chromogranin A (CgA) in sows with PDS. CgA is a stress marker of the sympathetic nervous system known to regulate vascular homeostasis. An increase in CgA can lead to sympathetic activation, leading to tachycardia and hypertension, as well as increased capillary wall permeability (Kaiser et al., 2020). This leads to increased vascular leakage which is central to oedema (Kaiser et al., 2020). This could also be an explanation for reduced colostrum quality in sows with SME. Before parturition, tight junctions of the capillary endothelium close in order to prevent milk components, for example, IgG, from the mammary gland alveoli to leak back to the bloodstream. This closure of tight junction may be impaired in sows with PDS.

Thus, determining the state of the cardiovascular system can be one way of diagnosing PDS. Besides an increase in heart rate (tachycardia), a prolonged capillary refill time in the vulvar mucosa (>2 s) has been reported by Kaiser et al. (2020). The degree of mammary gland oedema can be graded and SME can be diagnosed visually or via ultrasound (Björkman et al., 2017a, 2018a). At visual inspection, sows with severe udder oedema have dimpled skin with persistent marks on the floor. Further, teats are swollen and mammary glands are indistinct (Björkman et al., 2017a, 2018a). At palpation, glands feel doughy and pressure leaves an indentation (Kaiser et al., 2020). Ultrasound of the mammary glands shows thickened dermal and subdermal tissues, hyperechoic lobuloalveolar tissue with enlarged blood vessels and severe shadowing (Björkman et al., 2017a, 2018a; Björkman and Grahofer, 2020).

4 Infectious diseases causing fertility problems and reduced reproductive performance

From the physiological and immunological viewpoints, pregnancy in the pig can be divided into three phases. The first trimester involves recognition of the existence of embryos and the establishment of pregnancy, rendering the phase sensitive to environmental effects such as stress (see below), seasonality, especially light and temperature, and transmission of pathogens followed by early disruption of pregnancy. The middle phase is considered more stable, culminating in acquiring immune competence by the foetuses at around day 70 of pregnancy. Therefore, viral infections prior to this milestone can lead to mummification of embryos and prolonged length of pregnancy. The last trimester is characterized by foetal immunocompetence and therefore a safer phase that requires a high level of pathogenicity of the causative agent in order to be able to induce an abortion, as may happen with pathogens such as PRRS and *Leptospira* sp.

4.1 Viral causes for abortions

4.1.1 Porcine parvovirus (PPV) and picornavirus

In the scientific literature, porcine parvovirus (PPV) is a classic example of a viral pig disease that causes reproductive failure in the sow, which is manifested usually without visible abortion. PPV has also been used as an example of the temporal relationship between the acquisition of immunocompetence and the development of embryos and foetuses (Mengeling et al., 2000). During the embryonic period of development, infection is likely to cause death and resorption of developing early embryos. In the foetal period, if the infection occurs before day 70 of pregnancy, the foetuses lack the ability to respond and therefore become mummified. However, if infected after immunocompetence has been acquired, then the consequences are less serious, resulting in a slight weakness of the newborn piglets or, in some cases, no clinical signs at all.

PPV disease is globally widespread, with herd prevalence in most countries/ studies exceeding 70% (Nash, 1990; Oravainen et al., 2005). The primary clinical symptoms are reported to return to oestrus and birth of dead born, mummified foetuses. However, abortion storms have been described in conditions where naïve female herd faces the virus for the first time (Brown et al., 1980). Thus for most producers, it is a disease with which they have to cope, because eradication is not feasible. The control programme involves vaccination of all breeding animals in the herd: gilts, sows and boars. As PPV is known as the 'gilt disease', it is essential that gilts are vaccinated twice, before they are inseminated for the first time. Vaccine-induced immunity is boosted in a stepwise manner, so that sows that have farrowed at least once clearly have higher antibody titers than unbred gilts (Oravainen et al., 2005). Therefore, in practice, signs of PPV may be seen in cases where vaccination of replacement gilts has not been carried out, or as a result of a vaccination failure. Common vaccination failures can be due to the use of out-of-date vaccines, an inappropriate injection site (neck instead of the base of the ear), storage problems (frozen/overheating of vaccine during storage) and improper vaccination technique. It may also be that some animals turn out to be lacking the ability to respond because of the remaining maternally derived immunity (can last until 7 months for PPV) that inhibit vaccine responses

In addition to PPV, different genera of picornaviruses are known to cause various reproductive problems in pigs. Encephalomyocarditis virus (EMCV) has been found and isolated from a number of mammalian species (rodents regarded as the main host) and is known for reproductive pathogenicity with virus-excreting pigs being a risk for other pigs. Therefore, pig-to-pig contact between healthy and diseased or dead pigs is considered as the main route of transmission; however, transplacental transmission followed by abortions may occur. Presence of rodents is considered as a risk factor for the spread of EMCV, even in a piggery setting, serving as a reservoir for infection. High mortality in weaners after suffering from general systemic symptoms is described with myocardial failure/myocarditis as a specific clinical sign. In breeding animals, abortion, increased number of mummified and stillborn foetuses have been described as the main clinical findings (Zimmerman et al., 2019). Control of EMCV is based on strict rodent control in endemic areas. Furthermore, low-level exposure of the virus through manure to naïve pigs may trigger some immunity. Vaccination is also available, being efficient and of value for control in endemic areas.

Some strains of porcine enteric picornaviruses (porcine teschovirus, PTV, strains PTV-1,-3,-6) have been associated with reproductive symptoms such as abortion, stillbirth, mummified foetuses, embryonic death and infertility.

4.1.2 Porcine circo virus (PCV 2/3)

Porcine circo virus (PCV 2 and PCV3) infections have been reported to cause reproductive disease including abortions in sow herds in a number of countries, including South Korea and Spain (Chae, 2005; Maldonado et al., 2005; Saporiti et al., 2021). The virus is described as infecting the conceptuses following transplacental transmission. Embryonic or foetal death and loss of the whole litter may occur at any stage of pregnancy (Chae, 2005). In addition to abortion, an increased number of stillborn and mummified foetuses, and weak born piglets, may be observed (Segales, 2012; Table 1). The oro-nasal route of transmission is considered to be the most common route of infection; horizontal transmission between pen mates and from one pen to another may be very efficient. However, transmission through semen may also occur, although it has not yet been proven to be a significant route. Herds suffering from the epizootic phase of infection associated with reduced farrowing rates and increased stillborn rates may benefit from vaccination programmes that have been developed (Segales, 2012). PCV3, since its first recovery in 2015, has received considerable research attention. It is apparent that in addition to having the ability to cause systemic disease in growing pigs, it also can cause a reproductive disease as described, which however needs further research.

4.1.3 Porcine reproductive respiratory syndrome (PRRS)

PRRS occurs causing severe illness with pyrexia, inappetence, listlessness, bleeding and swelling of the ear and, as the name implies, respiratory

Table 1 Clinical manifestations, gross lesions, microscopic lesions and diagnosis in PCV2 reproductive disease (PCV2-RD)

Clinical manifestation	Gross lesions	Microscopic lesions	Diagnosis
Abortions	foetal mummification	non-suppurative to necrotizing or fibrous myocarditis of foetuses	1 late abortion 2 myocarditis 3 high/moderate PCV present in heart tissue by qPCR
	foetal hepatic enlargement		
	foetal cardiac hypertrophy	hepatic congestion	
	ascites in foetuses	pneumonia in foetuses	
Regular/ irregular returns to oestrus			1 regular/irregular return to oestrus PCV 2 2 seroconversion of PCV2 antibodies following the return to oestrus

Source: Modified from Segales (2012).

disease with bronchopneumonia and pleurisy in growing pigs. In gilts and sows, it affects the reproductive system and causes late-term interruption of pregnancy, mummified foetuses, stillbirths and weak born piglets (Hopper et al., 1992; Zimmerman et al., 2019). However, the transplacental infection may occasionally also occur earlier during pregnancy. During the acute phase of the disease, up to 1–3% of litters may be lost and mortality of sows is elevated by a few percentage points and is associated with pulmonary oedema and cystitis/nephritis. Late-term abortion occurs on days 100–118 and anything of the following may be encountered: normal piglets, a lot of variation in piglet birth weight, increased intrapartum death rate (piglets looking fresh), autolytic piglets and partially or completely mummified piglets (Zimmerman et al., 2019).

Routes for shedding of the virus include saliva, nasal secretions, urine and faeces. In semen, the virus may be shed for up to 43 days after exposure. It has also been reported that foetuses may die from hypoxia due to arteritis of the umbilical vessels (Lager and Halbur, 1996). Exposure of PRRS naïve replacement gilts to viral-infected tissues confirmed positive resulted in pre-breeding infection of gilts followed by sufficient immunity to protect the subsequent pregnancy (Menard et al., 2007). Lesions found in reproductive tissues include microscopic lesions in the uterus, oedematous myometrium and endometrium, with lymphohistiocytic perivascular cuffs. Prevention and control strategies involve internal and external biosecurity measures to maintain negative status in herds and areas that are free of the disease on one hand and keep new PRRSV variants out in those herds and areas that are endemic in terms of their PRRS status. In the latter situation, control is largely based on the application of internal biosecurity measures such as all-in all-out animal flow or partial depopulation. Application of vaccination programmes of breeding animals at adequate age to induce immunity prior to pregnancy belongs to the core of control measures in endemically infected herds. Eradication programmes based on total and partial depopulation/repopulation, test and removal have been described (Corzo et al., 2010). In addition, breeding for more resilient animals against PRRS has been recently proposed as an alternative strategy for the future (Knap and Doeschl-Wilson, 2021).

4.1.4 Aujzesky's disease (pseudorabies)

Another viral disease of pigs, mainly affecting the reproductive system, is Aujeszky's disease, due to infection with a herpes virus that may cause severe/ fatal disease in cattle and other species such as sheep, cats, dogs, mice and rats (Pensart and Kluge, 1989). Herd prevalence of pseudorabies in Western countries is usually low, in fact, most EU countries are free of the disease, but once infected, a herd with a continuous pig production system may suffer from it for a long time. The route of transmission of the virus is usually by inhalation or ingestion. It may also be transmitted by coitus, although there is some argument as to whether true venereal transmission occurs. Aujeszky's disease is characterized by nervous and respiratory signs, associated with a rise in temperature and often death in young piglets. Depending on the timing, infection in breeding adults may result in embryonic death, resorption of foetuses, abortion or stillbirth (Mettenleiter et al., 2019). In adult boars and sows, the clinical signs of this disease are seldom severe and usually consist of pyrexia, depression and anorexia that lasts for up to a week. Of great significance to the breeding herd is the fact that the virus causes embryonic death, foetal mummification and stillbirths.

4.1.5 Other viral agents

Several other viruses have been associated with reproductive disorders in the pig such as ASF and CSF (Almond et al., 2006). However, they are more likely to exert their effect indirectly on reproduction by causing systemic illness in the sow or gilt, such as severe pyrexia rather than attacking the reproductive system directly.

4.2 Bacterial infections

4.2.1 Leptospirosis

Leptospirosis is known to be a widely spread disease compromising reproduction in the pig population worldwide. *Leptospira interrogans serovar pomona* is the most widely recognized pathogen, but serovars *tarassovi*, *bratislava* and *icterohaemorrhagiae* may also be pathogenic in pigs (Arent and Ellis, 2019). Leptospira has a predilection site for the kidneys where they persist and multiply, and they are then secreted via the urine into the environment intermittently for a period of up to 2 years. Rodents are often thought to be the most significant reservoir species, maintaining the infection in the production environment. During the acute phase of the disease, mild general symptoms such as fever and anorexia may be seen; however, this phase often passes unrecognized. During the chronic phase of the infection, abortion rates of sows increase considerably, especially during the second half of gestation (Ellis, 2006; Arent and Ellis, 2019). Furthermore, stillbirth rates may increase and newborn piglets may lack viability (Nagy, 1993). Diagnosis is based on serological testing microscopic agglutination test, MAT and fluorescence antibody testing, FAT and ELISA and attempts to demonstrate the presence of leptospires in tissues and body fluids by qPCR and DNA sequencing (OIE, 2008; Ellis, 1990; Arent et al., 2016; Arent and Ellis, 2019). Prevention is based on rodent control and better hygiene, particularly, improved urine drainage from the floor surface. In addition, antimicrobials (tetracycline and streptomycin) can be used for treatment, and vaccines against specific serotypes are available for prevention.

4.2.2 Brucellosis

Brucellosis, due to infection with a number of *Brucella* species, has been recognized as a disease associated with abortion in the pig since the early 1900s. Although they may not be highly pathogenic for the pig, the zoonotic nature of the disease presents a risk for humans handling pigs, and interspecies transmission between other domestic animal species makes it an important infectious disease (Olsen et al., 2019). The most important *Brucella* species in the pig is *B. suis*; however, other species such as *B. neotomae, B. ovis, B. canis, B. abortus* and *B. melitensis* have also been identified (Olsen et al., 2019).

The most significant route of transmission is direct contact between animals, and infection is usually acquired through the oro-nasal or genital route. Adult pigs that are infected usually lack general systemic signs, such as fever and loss of appetite. Abortion may occur at any stage of pregnancy; however, other clinical signs relating to reproduction such as stillbirth and infertility may be more significant from the reproductive viewpoint (Olsen et al., 2019).

As sows usually eliminate the bacteria within 30 days after infection, a period of reproductive rest of two oestrous cycles duration (42 days) may be enough to ensure that the infection has been eliminated, and subsequent AI may be warranted. Other signs of the disease include subfertility, orchitis, posterior paralysis and lameness. Boars may be unable to eliminate the bacteria, and these may act as carriers within a herd. Antimicrobial therapy is often ineffective, and an eradication programme may offer the best outcome.

4.3 Other bacterial diseases and parasites

Chlamydia, specifically *C. abortus*, has been reported in a few studies in conjunction with abortion, weak neonates and irregular returns to oestrus in the pig (Camenisch et al., 2004; Broes et al., 2019). Bacteria causing an acute systemic response resulting in pyrexia present a risk to pregnant sows and gilts. For example, endotoxemia caused by gram-negative bacteria, or generalized, acute clinical signs caused by *Erysipelothrix rhusiopathiae* can cause abortion at any stage of pregnancy (Opriessnig and Coutinho, 2019). The better the immunity of the herd towards these pathogens, the lower the risk of reproductive disorders. Parasites may also cause reproductive problems including abortions. The connection between *Toxoplasma gonadii* and late-term abortions is well established (Lindsay et al., 2019). The cat serves as a definite host of the infection, whereas the pig serves as an intermediate host by contact with infected other intermediate hosts such as mice and rats (Lindsay et al., 2019). Infected pork is a source of Toxoplasma infection for companion animals and humans.

5 Non-infectious factors causing fertility problems and reduced reproductive performance

5.1 Group housing, stress and social interaction

In group housing, sows can interact with each other, which can lead to positive but also negative social interactions. Aggression between sows is the most prominent form of negative social interaction and can affect sows' health, welfare and reproduction (Spoolder and Vermeer, 2015). Group housing has been linked with a decrease in pregnancy rate, farrowing rate and litter size (Munsterhjelm et al., 2008). Whenever mixing sows, establishing a dominance hierarchy and competition over resources such as space and food will lead to acute and eventual chronic stress. Therefore, appropriate management of the animals, the feeding system, the size and design of the pen, group size and the skills and attitude of the stock person are important to maximize reproductive success (Spoolder and Vermeer, 2015).

It is well documented that the embryonic period of the pregnancy is more vulnerable for loss of pregnancy than the subsequent foetal period (Tast et al., 2002). Stress negatively affects embryo survival, in particular, during early pregnancy, around the period of attachment of the embryos to the uterine wall (days 11-16), and the period shortly thereafter when maternal recognition of pregnancy takes place (Spoolder and Vermeer, 2015). Thus, it seems wise to avoid stressful events in early pregnancy, especially in weeks 2–3 following insemination, to avoid a decrease in productivity. Recognition among gilts and sows appears to be based on familiarity gained when reared or housed together, and sows can recognize each other even after several weeks of separation (Spoolder et al., 1996; Stookey and Gonyou, 1998). Rank order fights are initiated when animals do not recognize each other or when they dispute each other's position in the hierarchy (Spoolder and Vermeer, 2015). The social hierarchy is usually established within 2 days after the introduction of new animals (Eliasson-Selling et al., 2000), but full integration can take at least 3–4 weeks (Moore et al., 1993; Spoolder and Vermeer, 2015). After social hierarchy has been established, aggressions usually decrease dramatically (Bokma et al., 1984; Luescher et al., 1990). Therefore, it is advisable that sows are mixed before they become pregnant and that sows know each other before mixing. Best practice would be to rear gilts together from an early age on and to keep familiar groups of gilts/sows in stable groups. This ensured that they can establish a social hierarchy long before they are inseminated for the first time. After lactation, it would be good to keep the sows in the same group and keep the group as stable as possible. Further, mixing of familiar sows should occur right after weaning. Most important is that the rank is established before fertilization takes place. This decreases the risk related to aggression during the embryonic period.

Another benefit of mixing sows right after weaning and before insemination is that social contacts between females are known to advance and induce the onset of oestrus, alleviating oestrus detection (Eliasson-Selling et al., 2000; Peltoniemi et al., 2016). Group-housed sows at AI can show oestrus freely compared to sows restricted to stalls (Eliasson-Selling, 2000) which may improve the possibilities for observing oestrus behaviour and reduce the risk of inseminating a sow that is not in standing oestrus (Einarsson et al., 2014). However, there is some evidence to suggest that social interactions may suppress oestrus signs in subordinate sows (Tsuma et al., 1996). It is advisable to separate individual subordinate sows or a small group of sows from the rest of the group for heat detection and artificial insemination. This can be done, that is, by taking the sow/small group of sows out of the pen and moving them to the boar.

Further, it should be noted that social interaction during the establishment of a dominance order (fighting) and/or during the exhibition of oestrus behaviour (mating) are temporary behaviours of sows and require more space per sow as compared to behaviour expressed by pregnant sows (Spoolder and Vermeer, 2015). All animals need space to rest, eat, defecate and explore the environment (static and behavioural space) but also to perform social behaviours (interaction space) (Spoolder and Vermeer, 2015). Grouping sows in a situation where there is insufficient space will increase aggressive behaviours and induce chronic stress (Barnett et al., 2001; Remience et al., 2008). The housing system should therefore be designed in a way that allows sows to show proper social behaviour, and that gives subordinate sows a possibility to escape from confrontation with another sow. It is also important to make sure that all sows can eat undisturbed. Regarding space allowance, a number of studies suggest that as a general rule sows should be provided with 2.5-3.5 m² per sow in order to reach a reasonably high farrowing rate (Kemp and Soede, 2012). In addition, feeding enough bulk and fibre and providing the pen with areas to escape help to alleviate aggressive behaviour problems of housing (Peltoniemi et al., 1999). Furthermore, it is regularly reported that having a teaser boar in the pregnant sow group calms the group down and reduces aggression among sows (Peltoniemi et al., 2016). For the group size, small groups may be more natural for sows. However, large dynamic groups of up to 300 sows may provide the farmer with satisfactory fertility results given that there is enough space in the group and if they are otherwise well managed.

Floor type in a pregnant sow unit appears as an important determinant of health (reviewed by Maes et al., 2016). Use of unbedded, slatted floors increases the incidence of lameness in group housing, probably due to claw trauma/infections rather than joint problems (Maes et al., 2016). Furthermore, in case of use of slats, use of concrete instead of plastic building material seems as an increased risk factor for lameness; however, plastic slats can be slippery. Slip-resistance, abrasiveness, surface profile and hardness are regarded as four important floor quality factors contributing to injuries of legs and claws (Maes et al., 2016).

5.2 Seasonal infertility

Seasonal infertility is defined as decreased fertility of the gilt and sow in summer and autumn. The signs of subfertility include: reduced pregnancy rate, prolonged weaning to oestrus interval (lower rate of oestrus expression) and delayed puberty of gilts (Fig. 1). In feral wild pigs, decreasing day length in late summer and autumn would provide a physiological cue, to indicate that it is not an optimum time for breeding, since the piglets would be born in the mid-winter. Thus in nature that would most likely mean death of the newborn piglets and loss of the whole litter. The feral wild pigs are seasonally anoestrous, with the onset of the anoestrous period being under the control of the photoperiod (Tast et al., 2001). However, the end of the anoestrus period and return to cyclical ovarian activity is dependent upon other environmental cues like the increased availability of food (Mauget, 1982). In the domesticated environment, cues such as increased availability of food and, in-housed animals, a comfortable microclimate are absent, and these cues for seasonal breeding are removed.

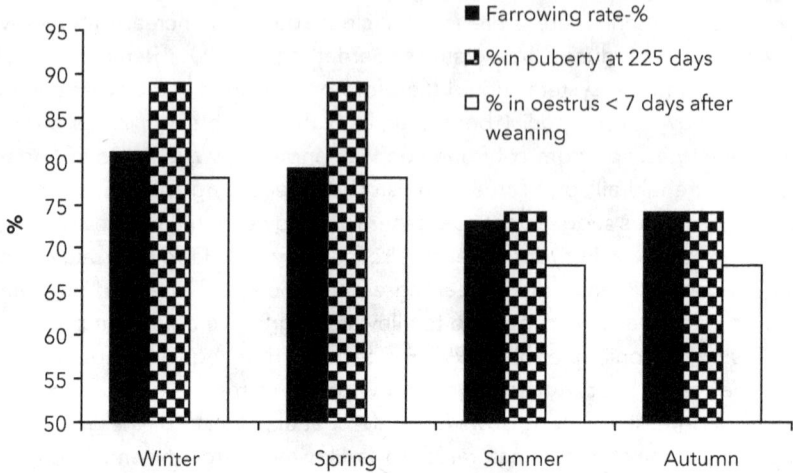

Figure 1 Major manifestations of seasonal infertility in the pig presented as percentages for each reproductive parameter. Data from Peltoniemi and Virolainen (2006).

Therefore, variable responses to decreasing daylight in the breeding activities of the female domestic pig population are frequently observed in late summer and autumn.

Seasonal infertility may be avoided by improving the reproductive management of the herd. This includes the use of a relevant light programme in which 16 h of light and 8 h of dark are provided in the breeding unit. This recommendation is combined with advice to explore and, if needed, increase the light intensity to a level of at least 200 lux at the level of the eyes of the sow. Feeding of pregnant, group-housed females during late summer and early autumn is also important. Increasing feed availability at this time is an effective method to decrease the stress of sows around the time of implantation of embryos and thereby alleviate early embryonic loss and return to oestrus. Moreover, the stress of gilts and sows prior to insemination and during pregnancy is reduced by grouping them correctly according to size and age. Furthermore, genetic selection, reduced stocking density in summer-autumn, cooling systems applied especially during lactation in a hot production environment and gonadotropin treatments may also be found useful in attempts to reduce seasonal effects on fertility (Peltoniemi and Virolainen, 2006).

Much of the reduction in farrowing rate in summer-autumn has been attributed to a disruption of early pregnancy rather than to conception failure (Love et al., 1993; Peltoniemi et al., 2000). These sows were mated, they conceived apparently well and pregnancy was established. The embryos were present for some time, but quite early on, pregnancy was interrupted and the sows returned to oestrus after prolonged oestrus to oestrus interval (Tast et al., 2002).

The diagnosis regarding seasonal disruption of pregnancy may be challenging since there is no diagnostic test for that. The diagnosis is by large based on the calendar, in other words, timing of the problem in late summer/ early autumn, and exclusion of other factors.

In the prevention of seasonal disruption of pregnancy, it is often beneficial to increase feeding for the duration of embryonic period of pregnancy (Peltoniemi et al., 1997). Other interventions that may prove beneficial include breaking down of the social group according to age and size, providing more space for the sows, including a boar in a group of pregnant sows, and taking care of lighting in the breeding unit (Peltoniemi and Virolainen, 2006; Auvigne et al., 2010).

5.3 Mycotoxins

Mycotoxins are fungal secondary metabolites produced by either field moulds (e.g. *Fusarium spp.*) or storage moulds (e.g. *Aspergillus spp.*) and *Penicillium spp.* (Eskola et al., 2020). In Europe, relevant mycotoxins are mainly derived from *Fusarium spp.* and include zearalenone (ZEA), trichothecenes (e.g. DON and T2) and Fumonisins (e.g. FB1 and FB2) (Eskola et al., 2020). Aflatoxins, a mycotoxin produced by the storage mould *Aspergillus spp.*, are usually abundant in warm and humid regions of the world, for example, in South America, Africa and South Asia (Eskola et al., 2020). Nevertheless, due to climate change, aflatoxins will also become one of the major mycotoxins gilts and sows are exposed to in Europe.

Mycotoxins can exert different effects depending on the production stage the gilts and sows are in. Exposure during rearing and sexual maturation can cause fertility problems, while exposure during gestation and lactation can also cause direct and indirect effects on foetal and neonatal development. Direct effects occur by transplacental and lactational transmission of mycotoxins from the dam to the foetuses and neonates (Dänicke et al., 2007; Goyarts et al., 2007; Sayyari et al., 2018; Benthem de Grave et al., 2021). Increased neonatal mortality, decreased neonatal vitality, reduced growth and impaired immune and organ functions are the results (Ensley and Radke, 2019; Kanora and Maes, 2010). Indirect effects occur through the impaired reproductive performance of exposed dams, for example, through reduced feed consumption, reduced colostrum and mild production, poor colostrum and milk quality, decreased maternal behaviour and increased restlessness and aggression (Ensley and Radke, 2019; Kanora and Maes, 2010). Thus, in gestating and lactating gilts and sows, mycotoxins can have various effects on the dams, their placentae and their foetuses and neonates depending on the type of mycotoxin, the dose and duration of exposure, and at what stage of gestation and lactation the exposure occurs.

The mechanisms of how mycotoxins exert their effects on sows' and gilts' reproduction as well as on foetal and neonatal development are not fully understood. It is known that mycotoxins have endocrine disruptive properties due to their oestrogenic activities interfering with the hypothalamic-pituitary-gonadal axis and impairing the blood-testes barrier in males, blood–milk barrier in dams and intestinal barrier in neonates (Ensley and Radke, 2019; Kanora and Maes, 2010). For instance, mycotoxins can lead to delayed maturity, pseudopregnancy, hyperoestrogenism syndrome, cystic ovarian degeneration, reduced litter size in female, delayed maturity and reduced testosterone, testis size and spermatogenesis in male. Further, they exert their toxicity at the molecular and genetic level (Ensley and Radke, 2019; Kanora and Maes, 2010). For instance, mycotoxins affect follicular growth, granulosa cell function and oocyte maturation in-vitro due to epigenetic modification, DNS damage and oxidative stress (Ensley and Radke, 2019; Kanora and Maes, 2010).

Mycotoxicosis is not easy to diagnose, but should be investigated whenever gilts, sows and boars show a reduction in their reproductive performance and piglets in their growth performance. Nevertheless, mycotoxicosis rarely causes a herd problem and a thorough anamnesis and history of the farm's reproductive management are important (Kanora and Maes, 2010). If mycotoxicosis is still suspected, the farmer needs to be advised on prevention, especially hygiene, and monitoring and control systems. Quality control programmes and regular check-ups of the quality of feed ingredients, including possible levels of mycotoxins present in the feedstuff are the base for prevention (Kanora and Maes, 2010). Further, advise on managing the critical conditions that facilitate the growth of fungi (e.g. dust, hygiene, temperature and moisture) in silos and feeders needs to be given and regular cleaning and disinfection needs to be encouraged (Kanora and Maes, 2010). Besides preventive measures, modern monitoring systems (Lawlor and Lynch, 2001; Berthiller et al., 2007) and other tools can be used for control. The other tools include toxin binders, acidifiers and dis-activators of the different mycotoxins present in feed and or feedstuffs (Dänicke et al., 2004; Völkl et al., 2004).

5.4 Lactation weight loss and metabolic condition at breeding

Management of sows' and gilts' metabolic conditions before breeding is important for their subsequent reproductive performance and piglets' growth performance after subsequent parturition (Soede and Kemp, 2015). For gilts, it is important that gilt development is optimized and that breeding occurs when gilts have reached an optimal body weight of 145–160 kg and an optimal body condition of 12-18 mm back-fat depth (Rozeboom, 2015). For sows, it is important that the energy and protein loss during previous lactation and the metabolic state at weaning are optimized because the following weaning-to-oestrus interval is too short to correct for a severe negative energy balance and to improve their subsequent reproductive performance (Kemp et al., 2018). The mechanism of action how the negative energy balance is affecting the next generation of growing follicles and thereby embryonic development appears IGF-1-dependent (Han et al., 2020, 2021).

During lactation, both the suckling intensity of piglets and the negative energy and protein balance inhibit follicles from growing. If lactation energy and protein loss are severe, follicle development is suppressed to such an extent that subsequent fertility of sows is impaired. This can result in negative effects on both the sows' reproductive performance and the piglet's growth performance. Sows can have a decrease in estrus rate, ovulation rate, pregnancy rate or litter size (Soede and Kemp, 2015). Lactation energy and protein loss are reflected in their body weight loss duration lactation. In multiparous sows, around 10-12% of body weight loss can be considered severe (Thaker and Bilkei, 2005), whereas in primiparous sows even bodyweight losses of less than 10% can be considered severe (Hoving et al., 2012). This is because young sows have higher energy requirement at breeding, not only for their maintenance and reproductive performance but also for their own body growth (Hoving et al., 2012). Further, primiparous sows have lower capabilities to meet these energy demands because they have a lower feed intake compared to older sows (Hoving et al., 2011).

The post-weaning follicular period is very short. Sows' metabolic condition needs to be optimized before weaning by preventing moderate (>5%) lactation weight loss in primiparous sows and severe (>10%) lactation weight loss in multiparous sows. This can be achieved by optimizing feed composition and intake during gestation, transition and lactation as well as by providing a stress-free and animal-friendly environment (Soede and Kemp, 2015). Another option is to reduce the suckling stimulus and thereby decrease the metabolic burden, for example, by split weaning or intermittent suckling (Soede and Kemp, 2015). Nevertheless, the benefit of decreasing the suckling stimulus on the metabolic state is rather marginal. The same applies to improving follicular development post-weaning by means of optimizing feed intake and composition during the weaning-to-oestrus interval because of its short duration. The only feasible management strategy would be to extend the weaning-to-oestrus interval by skipping a heat or by postponing the oestrus using a progesterone analogue for at least 10 days (Soede and Kemp, 2015). These management strategies should be considered in primiparous sows with moderate weight loss and in multiparous sows with severe weight.

Another possibility for those sows is to extend the lactation length up to 5-6 weeks. After the first 2-4 weeks of lactation, sows get into a positive energy balance as feed allowance reaches the maximum (Feyera and Theil, 2017). However, this positive energy balance in the remaining lactation may be too short to compensate for the body condition loss during lactation in those sows (Eissen et al., 2000). Allowing them 1–2 more weeks of lactation with maximum feed allowance may get them into better metabolic shape at weaning. Extending the lactation length can be combined with reducing the metabolic burden, for example, split weaning and intermittent suckling. If possible, special diets, for example, as dietary supplements/top-dressings, that stimulate IGF-1 production could be provided during this additional time (Han et al., 2020, 2021). Other factors that can affect post-weaning follicular development and subsequent reproductive performance are stress, social interactions between sows, boar contact, quality of oestrus detection and timing of insemination (Soede and Kemp, 2015).

6 Parturition

6.1 Physiology of farrowing

Physiology of farrowing is very complex, and several hormones act and interact to regulate the farrowing process. In the latest stages of pregnancy, progesterone, LH, oestrogens, cortisol, prolactin, relaxin and prostaglandins become the main actors that regulate all the physical events leading to parturition. All these hormones, regulated by various internal clocks, interact and influence each other in a very refined manner (Anderson, 2000). Behavioural expression is also very important for the farrowing process. Thorsen et al., 2017 Sows express an innate hormone-driven nest-building behaviour that signals the beginning of parturition (Algers and Uvnäs-Moberg, 2007). However, the environment can also have an important, though indirect, influence on the pattern of these hormones. A restricted environment can also prevent this nest-building behaviour, especially in the modern swine production system (Damm et al., 2003). This refined, delicate hormonal process clearly requires very well-defined premises in order to develop properly. In addition to housing, nutrition and disease may also affect, either directly or indirectly, the release or activity of farrowing hormones that interfere with the physiological process.

6.2 Hormones in pregnancy and parturition

During early pregnancy, after implantation of the embryos and the development activity of the corpora lutea, a higher and constant level of progesterone dominates the hormonal pattern (Meulen et al., 1988). This preserves the progress of pregnancy, and for almost two-thirds of its length, the level of progesterone circulating in the blood will remain high. After the first trimester of pregnancy, most of the other reproductive hormones (oxytocin, prostaglandins, relaxin and prolactin) show very basal levels or very little pulsating activity (Fig. 2A) for the remainder of the pregnancy. This very stable hormonal activity changes completely 48-24 h before the beginning of parturition. The quick drop in progesterone concentration produces a cascade effect on almost all the other hormones, which throughout the pregnancy remained quite stable. The level of prostaglandin peaks, the oxytocin concentration increases and begins to exhibit high pulsating activity (Gilbert et al., 1994), the prolactin concentration gradually increases, and oestrogens, after peaking quickly, gradually drop to basal levels (Ellendorff et al., 1979; Kindahl et al., 1982; Anderson, 2000) (Fig. 2B).

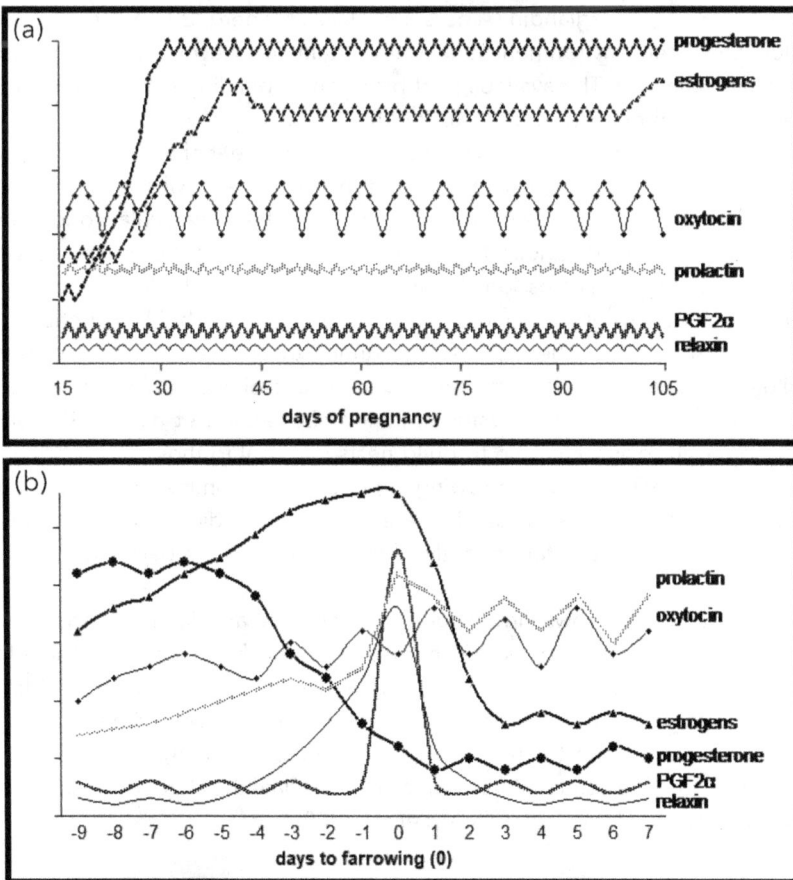

Figure 2 A schematic description of the level of reproductive hormones during pregnancy in the sow (modified from Anderson, 2000). Soon after early pregnancy, reproductive hormones show very stable levels for almost the entire pregnancy (a). Hormone levels undergo major changes only a few days before farrowing (b).

It is remarkable how such large hormonal changes occur in such a very limited amount of time.

6.3 Parturition behaviour and activity

Before and during farrowing, the intense hormonal activity described above is responsible not only for inducing the parturition event but also for triggering visible behavioural changes. Activities such as rooting, pawing, turning and walking increase considerably 24 h prior to farrowing (Hartsock and Barczewski, 1997; Oliviero et al., 2008b) and characterize the nest-building behaviour. One of the triggering factors of nest-building behaviour has been found to be a rise in prolactin (Castren et al., 1994), induced by a decrease in progesterone and an increase in prostaglandin (Algers and Uvnäs-Moberg, 2007). The external influence of the environment is also important for the expression of nest-building behaviour. The availability of proper nest-building material seems to speed up this process (Damm et al., 2000).

Most housing systems are based on the confinement of the farrowing sow in crates, where the sow has very limited movement and where bedding or any other nest-building substrate is often absent. In these particular conditions, the nest-building behaviour triggered by endogenous hormonal activity cannot find proper expression. In the absence of a nest-building substrate, confined sows express prolonged and unsuccessful nest-building behaviour (Damm et al., 2003). The lack of opportunity to express appropriate nest-building behaviour can lead to an increase in cortisol and ACTH (Jarvis et al., 1997), which indicates a stressful condition. Gustafsson et al. (1999) found that domestic sows were able to build nests identical to those of wild boars, even after several previous farrowing experiences in confined crates without bedding. This innate behaviour is therefore a clear indicator of impending farrowing and occurs independently of the housing or the bedding material available.

The benefits for sows farrowing in an open crate are significant and show how the welfare of the animals often correlates positively with economic benefit, as the lower number of stillborn piglets demonstrates (Oliviero, 2010). Adding roughage some days before farrowing would provide the sow with a substrate to better express its nest-building behaviour (Yun et al., 2014), and in the case of straw, a possible source of fibre which can alleviate the state of constipation that arises around farrowing (Oliviero et al., 2009).

6.4 Duration of farrowing and possible complications

As discussed in the introduction, the duration of farrowing has profoundly increased while the litter size has doubled over the past three decades (Fig. 3).

Litter size and farrowing duration

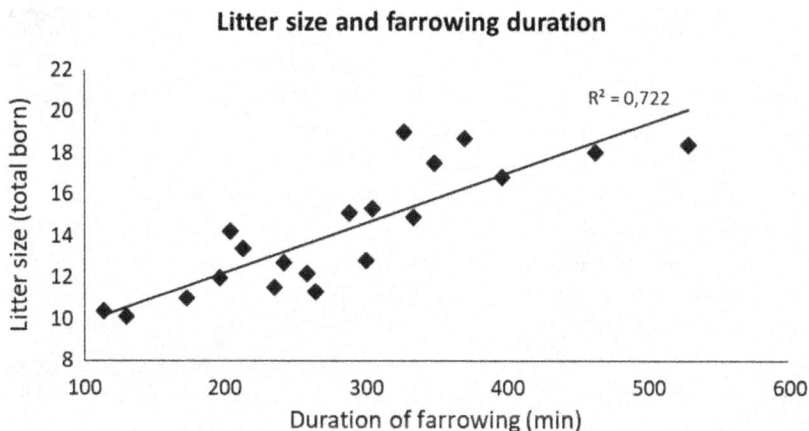

Figure 3 Relationship between litter size and the average duration of farrowing in 20 studies from 1992 to 2018 (adapted from Oliviero et al., 2019, Reproduction in Domestic Animals, Wiley-Blackwell).

The increasing litter size presents an immunological challenge for the sow and especially the piglets (Oliviero et al., 2013, 2019). The last 20–30% of the foetuses to be born likely miss out on access to good-quality colostrum that declines by 50% already by the sixth hour after the birth of the first piglet (Le Dividich et al., 2017). On the other hand, they also have less time to suckle colostrum due to decreased window of opportunity for colostrum intake, increased competition for teats and reduced birth weight. This all may show up later in the emergence of diseases in the growing phase of piglets/fattening pigs.

The metabolic challenge related to hyper-prolific sow production model begins in the growing phase of gilts and goes beyond farrowing and lactation. The sow is supposed to eat enough to meet the requirement of growing litters prior to farrowing, which may cause some of the problems seen around farrowing (Oliviero et al., 2008a, 2009). In the early part of lactation, sows with large litters loose more energy while producing milk than what they can consume, ending up in a negative energy balance (NEB) (Hoving et al., 2012; Costermans et al., 2020).

The growing litter size and intensity of production have triggered welfare concerns to the public. This seems to happen regardless of whether those concerns would be warranted or not.

7 Low state systemic inflammation involved in parturition and postpartum dysgalactia syndrome

Nest building and the phases of farrowing are orchestrated by responding to changes in reproductive hormones. It is well established that a decline in progesterone and peak in prostaglandin F2alpha triggers nest-building behaviour while oxytocin rise at the beginning of expulsion phase marks the session of nest building (Algers and Uvnäs-Moberg, 2007). Prostaglandin F2alpha peak also induces CL regression with a concomitant decline in progesterone, making uterine contractions and parturition possible. Oxytocin is mainly in charge of uterine contractions during the expulsion phase of parturition and letdown of colostrum and milk, while prolactin will promote mammary gland development to the extent that initiation of milk production after parturition will become possible (Taverne and van der Weijden, 2008; Farmer and Quesnel, 2009; Farmer, 2016).

It has also been described in the literature and also shown by our group that allowing the sow to build up a nest prior to farrowing will increase oxytocin release and shorten the duration of farrowing (Castren et al., 1993; Oliviero et al., 2008a). Other ways of shortening the duration of farrowing include increasing fibre in the feedstuff and encouraging water intake (Oliviero et al., 2008a, 2010). However, even after applying the best management interventions prior to farrowing, the duration of farrowing of modern hyper-prolific sows is extended four to five hold as described (Kaiser et al., 2018a; Oliviero et al., 2019; Yun et al., 2019).

Prolonged farrowing increases the risk for reduced quality and quantity of colostrum intake, intrapartum hypoxia of foetuses (Peltoniemi et al., 2020), retained placentae (Björkman et al., 2017b), uterine inflammation and PDS (Björkman et al., 2018b) and, likely, reduced development of next generation of follicles fertility (Oliviero et al., 2013; Peltoniemi et al., 2020).

Moreover, during the periparturient period, biological mechanisms coordinate the mobilization of body reserves in order to support foetal growth and milk production; insulin concentrations are reduced, and the response of hormone-sensitive lipase in adipose tissue (e.g. low insulin, high growth hormone and catecholamines, or high glucocorticoid concentrations) is greater to facilitate lipid mobilization. This periparturient period is also characterized by a low state of inflammation encompassing an increase in hepatic production of positive acute-phase proteins (APP) and a decrease in the production of negative APP (Petersen et al., 2004; Kaiser et al., 2018a). The evolving low state periparturient inflammation may be underway to be more profound in the hyper-prolific sows as litter size increases (Kaiser et al., 2018a). Also, the use of non-steroidal anti-inflammatory drugs (NSAIDs) seems useful in the control of periparturient inflammatory process like PDS (Schoos et al., 2020). However, this may be elucidated in further studies in the future.

It has been rather well described in the literature that these responses are mediated by the pro-inflammatory cytokines interleukin (IL)-6, IL-1β, and tumour necrosis factor-α (TNF-α) (Kaiser et al., 2018a). Additionally, evidence in the dairy cow indicates that oxidative stress also occurs during this period and is driven by the imbalance between the production of reactive oxygen metabolites (ROM), reactive nitrogen species (RNS) and the neutralizing capacity of antioxidant mechanisms in tissues and blood (Coleman et al., 2020).

The extent and duration of the inflammatory process will determine whether or not the condition is ending up as a clinical disease. However, it is noteworthy that in the hyper-prolific sows lines as those typical of Denmark and Belgium, the incidence of sows contracting a systemic disease postpartum is as high as >30% (Larsen and Thorup, 2006; Papadopoulos et al., 2010). Moreover, it is obvious that even in those sows staying healthy as far as clinical symptoms, the inflammatory process is heavily present as indicated by means of those markers described above (Kaiser et al., 2018a).

Typically, within 2–3 days postpartum, the process of inflammation may develop into endotoxemia, which involves the release of the inflammation markers described. Endotoxemia is associated with clinical symptoms indicating a systemic response to infectious agents such as coliform bacteria – and PDS (Bäckström et al., 1984; Peltoniemi et al., 2016; Kemper, 2020). The condition comes with acute general symptoms such as inappetite, lethargy and fever (Peltoniemi et al., 2016), followed by local symptoms that usually affect either the uterus (Björkman et al., 2018b) or the udder (Farmer et al., 2019) or both of them.

After parturition, concomitant with the process of inflammation, the sow undergoes metabolic stress due to loss of body reserves in favour of milk produced for large litters. This change rate is highest during the first 10 days of lactation. One of the major mediators of metabolic stress is IGF-1, which is also seen as an indicator of fertility. Low IGF-1 levels are associated with the increment of inflammation, presence of metabolic stress and reduced fertility. IGF-1 is also regarded as one of the most important factors driving follicle development (Han et al., 2019, 2020). The role of extracellular vesicles, although proposed as being key players in follicle development and the cross-talk between the mother and the embryo, in this inflammatory process and its effect on follicle development, however, remains less explored (Almiñana et al., 2017).

In summary, in hyper-prolific sows, the physiological process of farrowing is prolonged, making the system vulnerable in terms of increased rate of inflammation and emerging infectious uterine and mammary disease. In fact, recent evidence shows that even in sows staying without symptoms, there seems to be a considerable degree of 'silent inflammation' in the body. In an increased proportion of sows, however, postpartum disease PDS is detected and hopefully treated. The consequences of inflammation, regardless of clinical symptoms, include reduced quantity and quality of piglet colostrum intake and milk intake during early lactation.

8 Transfer of immunity

At birth, piglets have no circulating immunoglobulin from the mother, due to the epitheliochorial placenta, not allowing transfer of such large-sized molecules during the foetal period. Therefore, piglets need to acquire immunoglobulin passively from colostrum to gain adequate immune protection, before they will be able to start producing their own immunoglobulin from 3-4 weeks of age onwards (Rooke and Bland, 2002; Alexopoulos et al., 2018).

8.1 Emergence of large litters creates a challenge for transfer of immunity

As discussed previously, the average litter size has increased from 11 piglets to 14 piglets, with peaks of 16 piglets in some countries (Kemp et al., 2018; Baumgartner, 2020). With hyper-prolific sow lines, however, litters up to 18-20 piglets have become common (Björkman et al., 2017b; Kemp et al., 2018; Schoos et al., 2020). This poses a direct challenge during lactation, since most sows have an average of only 14-16 teats (Labroue et al., 2001). In a study with no litter balancing nor providing any kind of direct help to sow and piglets, it was found that a sow can wean successfully no more than 10-11 piglets, independent of how large the litter was (Andersen et al., 2011). Large litters also have a direct effect on the piglets' characteristics at birth. The larger the litter size, the lower is the piglets' average birthweight and the higher is the within litter weight variation (Quesnel et al., 2008; Beaulieu et al., 2010; Smit et al., 2013; Matheson et al., 2018). Competition for colostrum intake is increased with more piglets born than the available teats at the sow's udder, further increasing the weight variation within the litter (Vasdal and Andersen, 2012; Declerck et al., 2017). Consequently, being born in a large litter with a low birth weight decreases the piglet vitality, which can retard the access to the udder for adequate colostrum intake (Hoy et al., 1994; Islas-Fabila et al., 2018; Manjarin et al., 2018). Colostrum intake significantly influences piglets' performance as well as their survival, not only prior to weaning but also in the long term. As colostrum yield is reported to be independent of litter size, sufficient colostrum intake per piglet is crucial, especially in hyper-prolific sow (Declerck et al., 2016; Balzani et al, 2018).

Colostrum yield and its fat content are largely affected by different sow and litter factors. Pig producers should consider colostrum intake, yield and composition in their management to maximize production and reproduction potential (Declerck et al., 2015).

8.2 Transfer of local immunity in the gastrointestinal tract and gut closure

Sow's milk has a constant presence of secretory IgA (sIgA) which ensures good protection of the piglets' gut. This localized protection allows piglets to develop gradually their own immune response mechanisms, as long as they are able to suckle sufficient milk (Salmon et al., 2009). Other types of immunoglobulin, like IgG, are abundant in colostrum (Carbera et al., 2013; Theil et al., 2014). Sow colostrum has a high content of IgG (30-70g/l) and other bioactive compounds like growth factors and enzymes (Oliviero et al., 2019).

One limiting factor for piglets on the absorption of IgG from colostrum is the closure of the tight junctions in the gut occurring 24-36 h after birth, making piglets unable to absorb such large molecules. For this reason, the main cause of death in newborn piglets is the impossibility to obtain a sufficient amount of colostrum (Quesnel et al., 2012). A minimum of 250 g of colostrum is considered the minimum requirement to reduce mortality and allow proper growth (Quesnel et al., 2012).

8.3 Colostrum quantity and challenges due to competition for functional teats

Another limiting factor is the ability of the sow to produce adequate amount of colostrum to sufficiently supply large litters. According to Lessard et al. (2018) birth weight and colostrum intake can influence the genes' expression of immunity and oxidative stress in piglets' intestinal tissue. Low birth weight piglets had less intestinal antigen-presenting cells and altered development of B cells, when compared to piglets with higher birth weight (Lessard et al., 2018).

In large litters, immunity is also affected by social stress conditions (competition for colostrum and milk intake, crowding and regrouping). Psychosocial stress may alter both innate and adaptive immune responses, such as leukocyte distribution, cytokine secretion, lymphocyte proliferation, antibody production and immune responses to viral infection or vaccination (Gimsa et al., 2018). Additionally, social stress may lead to increased cortisol levels, promoting dysregulation of inflammatory processes and glucocorticoid resistance of lymphocytes and therefore predispose to gastrointestinal diseases (Gimsa et al., 2018).

There is an association between large litter size and increased pre-weaning mortality (Baxter et al., 2013; Rutherford et al., 2013), of which one visual example is shown in Fig. 4. Prolonged farrowing duration and low birth weight seen in large litters can explain the causes of this correlation (Oliviero et al., 2019).

8.4 Problems evolving due to prolonged process of farrowing

Prolonged parturition due to large litters can affect the level of absorbed IgG in piglet plasma with a linear decrease of 0.4 g/L for each piglet born (Kielland

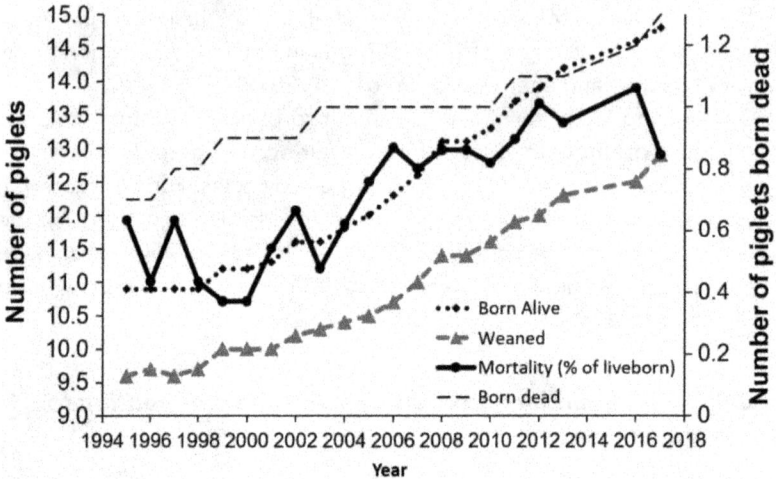

Figure 4 Increased mortality with increased litter size in the Netherlands (adapted from AgroVision B.V. The Netherlands, 2017).

et al., 2015). Many studies report a negative correlation between litter size and piglet birth weight (Quesnel et al., 2008; Beaulieu et al., 2010; Kapell et al., 2011; Smit et al., 2013). A decline in average birth weight relates to large litters with 35–43 g lower birth weight for each additional pig born (Quiniou et al., 2002; Beaulieu et al., 2010; Smit et al., 2013). The lower birth weight directly reduces colostrum intake, increasing the risk of mortality (Ferrari et al., 2014; Declerck et al., 2016; Le Dividich et al., 2017).

Piglets born with higher birth weight had higher serum IgG concentrations (Nguyen et al., 2013). Similarly, higher colostrum intake at birth increased the IgG content in serum of piglets at 24 h after birth (Moreira et al., 2017) and also at 10 and 20 days of age (Ferrari et al., 2014).

Immunoglobulin content in colostrum declines rapidly after the first piglets are born (Le Dividich et al., 2017), therefore prolonged duration of farrowing and the higher competition in large litters can reduce not only the possibility to suckle sufficient colostrum but also retard the time of access to the udder. Delaying the intake of colostrum affected negatively the piglets' immunoglobulin absorption and the maturation of their intestinal villi, with possible long-term detrimental effects on the digestion process (Cabrera et al., 2012). Due to the rapid decrease of IgG content in colostrum, Klobasa et al. (1987) found that birth order affected the amount of immunoglobulin absorbed by piglets. The latest piglets born in the litters had the lower plasma IgG level. Correspondingly, a 4% decrease in plasma IgG concentration was found in piglets of smaller birth weight (Cabrera et al., 2012). It is clear that the farrowing-to-suckling interval is fundamental in the acquisition of adequate IgG by piglets. After the start of parturition, a delay of 4 h in up taking of colostrum reduced the amount of plasma proteins from 24 h up to 12 days in piglets. It is therefore extremely important to consider also the time of birth of piglets in relation to the start of farrowing, when planning successful strategies to boost colostrum intake in large litters, like for instance split suckling (Oliviero et al., 2019).

9 Microbiota involvement during pregnancy, parturition and lactation

The gut microbiota has a constant shift over time, and in sows, both diversity and abundance of gut microbial population increase from pregnancy until the end of lactation (Ji et al., 2019). A more diverse gut microbiota can provide a more complete metabolism and functionality in sows, allowing a sufficient supply of nutrients for foetal growth and development (Ji et al., 2019). A recent study found that at farrowing, the most predominant phyla were Firmicutes, Bacteroidetes, Proteobacteria, Actinobacteria and Candidatus (Hasan et al., 2018a). The Firmicutes represent the most abundant proportion of the total population, followed by Bacteroides.

These two phyla accounted for approximately 98% of all bacteria present. These results are in line with the one published by Kim et al. (2011). However, Bacteroides increase linearly from the start to the end of pregnancy (Ji et al., 2019). Jost et al. (2014) reported that Firmicutes exhibited no detectable changes over perinatal period.

It seems that in sows, the increase in body weight or in the back-fat thickness in the different production phases is associated with the higher abundance of Firmicutes or the Firmicutes to Bacteroides ratio (Kim et al., 2011; Ji et al., 2019). In terms of phyla, the abundance of Tenericutes, Fibrobacteres and Cyanobacteria has been shown to increase with the progression of the pregnancy (Ji et al., 2019). Tenericutes have been connected with better intestinal cells' integrity and Fibrobacteres by having the potential to metabolize non-soluble polysaccharides, such as cellulose, hemicellulose or pectin (Ji et al., 2019).

In sows at late gestation, Romboutsia was the dominant genus, followed by Clostridium *sensu stricto*, Lactobacillus, Oscillibacter, Intestinimonas, Sporobacter, Christensenella, Barnesiella, Flavonifractor, Terrisporobacter, Acidaminobacter, Lachnospiracea incertae sedis and Turicibacter, with other genera being much less than 1% (Hasan et al., 2018a).

Another important factor responsible to modulate gut microbiota composition is diet, possibly having functional effects. Functional ingredients added to the diet, like hydrolysed yeast (Hasan et al., 2018a) and resin acid-enriched composition (Hasan et al., 2018b), probiotics (Menegat et al., 2019) and prebiotics (Tan et al., 2016; Li et al., 2020), showed the ability to modify the microbial gut population in sows. Other compounds of the diet like protein and fibre can modulate the gut microbial population both in gestating sows and in weaning piglets. Fibres can be classified as insoluble (like whole-wheat flour and wheat bran) and soluble (oats and barley). Fibre showed the most interesting properties when provided during late pregnancy, reducing Proteobacteria and increasing Ruminococcaceae, Oscillospira and Eubacterium (Tan et al., 2016; Li et al., 2020).

After dietary soluble fibre supplementation in late pregnancy sows, the genus Eubacterium increases, promoting propionate release (Xu et al., 2020). These microbiota are capable to ferment indigestible carbohydrates, producing short-chain fatty acids (SCFA) that can be an important energy source for the sow. Butyrate is considered to have gut health-promoting features, acting as the main energy source for colonocytes and having anti-inflammatory properties at gut level (Sassone-Corsi and Raffatellu, 2015).

With an increased production of SCFAs, there is more energy available at intestinal level, contributing to the high energetic demands of hyper-prolific sows. Therefore, in gestating sows, it could be favourable to promote fibre degrading microbiota in the gut. When dietary supplementation can modify substrate availability and gut physiological processes (fermentation, pH), due to the increase of beneficial bacteria, this also implies a reduction in pathogenic bacteria (Liu et al., 2008). After supplementation of yeast hydrolysate to sows during pregnancy, beneficial and fermentative bacteria (Roseburia, Paraprevotella and Eubacterium) increased, while some opportunistic pathogens like Desulfovibrio, Escherichia/Shigella and Helicobacter, belonging to the phylum Proteobacteria, were reduced (Hasan et al., 2018a).

Proteobacteria are usually not very abundant within a normal gut microbial community of adult animals (Eckburg et al., 2005). However, their expansion may indicate a dysbiosis in connection with gut inflammation (Shin et al., 2015; Litvak et al., 2017), like irritable bowel syndrome or inflammatory bowel disease (Morgan et al., 2012), and with increased inflammatory responses of women in late pregnancy (Koren et al., 2012).

For instance, Hasan et al. (2018a) found that some positive sow's productive and physiological performances (high colostrum yield, high colostrum proteins content, high colostrum IgG content, normal blood progesterone level and normal farrowing duration) were positively correlated to the gut bacterial families Lactobacillaceae, Ruminococcaceae and Prevotellaceae, the last two being bacteria capable to metabolize indigestible polysaccharides from dietary fibre.

Conversely, unfavourable productive and physiological performances of the sow (low colostrum yield, low colostrum proteins content, low colostrum

IgG content, high level of blood progesterone and long farrowing duration) were positively correlated with the gut bacterial families *Erysipelotrichaceae*, *Clostridiaceae*, *Streptococcaceae*, *Enterobacteriaceae*, *Desulfovibrionaceae* and *Bacteroidaceae*, some of these being pig pathogens bacteria or part of the phylum Proteobacteria.

10 Conclusion and clinical implications

Health challenges during pregnancy and parturition in the pig include a variety of internal and external factors that threaten the general health of the gilt and sow on the one hand and specific reproductive processes on the other hand. Physical quantities of the production environment, including spacing, flooring, feeding systems and the use of bedding materials provide the gilt and sow with a variety of physical environments that by large define possibilities for a healthy life. Internal and external farm and biosecurity determine how vulnerable gilts and sows are for pathogens threatening pregnancy and parturition. Season and chronic social stress in sow groups mainly disrupt pregnancies during the first half of pregnancy while abortions occurring during the last trimester of pregnancy more often point to a viral or bacterial cause. We conclude that breeding for ever-increasing litter size has prolonged the duration of farrowing by four to five fold over the past three decades, creating problems for transfer of immunity and thereby the health of the piglets. Mechanisms involved are numerous, including at least hormonal, behavioural and immunological alterations to what has been considered as normal and healthy. Transfer of immunity is affected by changes in microbiota that seem to be coupled with management of parturition and carry over to the general health of growing piglets. Breeding for resilience regarding health and welfare has been recommended for future studies and increased resilience against specific pathogens, and environmental changes may indeed pay off in the long term.

11 Where to look for further information

For online search engines, we recommend using keywords included in the subtitles of the present chapter. Key text books covering the areas discussed would include Diseases of Swine (the 11th edition, 2019, i.e. Neumann et al. 2019 in the list of references), Veterinary Reproduction and Obstetrics (the 10th edition, 2019, edited by Noakes, Parkinson and England, Elsevier) and The Gestating and Lactating sow (edited by Farmer, 2015) by Wageningen Academic Publishers. Regarding seasonal infertility, we recommend a case study in a Spanish herd, available online at: https://www.pig333.com/articles/diagnostic-darkness-recurring-fertility-decline-in-summer_14574/.

12 References

Alarcon, L. V., Allepuz, A. and Mateu, E. (2021). Biosecurity in pig farms: A review. *Porcine Health Management* 7(1):5. DOI: 10.1186/s40813-020-00181-z.

Alexopoulos, J. G., Lines, D. S., Hallett, S. and Plush, K. J. (2018). A review of success factors for piglet fostering in lactation. *Animals: An Open Access Journal From MDPI* 8(3):38. DOI: 10.3390/ani8030038.

Algers, B. and Uvnäs-Moberg, K. (2007). Maternal behavior in pigs. *Hormones and Behavior* 52(1):78-85.

Almiñana, C., Corbin, E., Tsikis, G., Alcântara-Neto, A. S., Labas, V., Reynaud, K., Galio, L., Uzbekov, R., Garanina, A. S., Druart, X. and Mermillod, P. (2017). Oviduct extracellular vesicles protein content and their role during oviduct–embryo cross-talk. *Reproduction* 154(3):153-168. DOI: 10.1530/REP-17-0054.

Almond, G. W., Flowers, W. L., Batista, L. and D'Allaire, S. (2006). Diseases of the reproductive system. In: Straw, B., Zimmermann, J. J., D'Allaire, S. and Taylor, D. J. (Eds) *Diseases of Swine* (9th edn.). Ames, IA: Blackwell Publishing, pp. 113-147.

Althouse, G. C., Kauffold, J. and Rossow, S. (2019). Diseases of the reproductive system. *Diseases of Swine*, pp. 373-392.

Andersen, I. L., Nevdal, E. and Bøe, K. E. (2011). Maternal investment, sibling competition, and offspring survival with increasing litter size and parity in pigs (Sus scrofa). *Behavioral Ecology and Sociobiology* 65(6):1159-1167.

Anderson, L. L. (2000). Reproductive cycle of pigs. In: Hafez, E. S. E. and Hafez, B. (Eds) *Reproduction in Farmanimals* (VII edn.). Lippincott, pp. 189-190.

Arent, Z., Frizzell, C., Gilmore, C., Allen, A. and Ellis, W. A. (2016). Leptospira interrogans serovars Bratislava and Muenchen animal infections: Implications for epidemiology and control. *Veterinary Microbiology* 190:19-26.

Arent, Z. J. and Ellis, W. A. (2019). Leptospirosis. In: Zimmerman, J. J., Karriker, L. A., Ramirez, A., Schwartz, K. J., Stevenson, G. W. and Zhang, J. (Eds) *Diseases of Swine* (11th edn.). Glasgow, Great Britain: Wiley Blackwell, pp. 854-862.

Auvigne, V., Leneveu, P., Jehannin, C., Peltoniemi, O. and Sallé, E. (2010). Seasonal infertility in sows: An observational field study to analyze relative roles of heat stress and photoperiod. *Theriogenology* 74(1):60-66.

Bäckström, L., Morkoc, A. C., Connor, J., Larson, R. and Price, W. (1984). Clinical study of mastitis-metritis-agalactia in sows in Illinois. *Journal of the American Veterinary Medical Association* 185(1):70-73.

Baer, C. and Bilkei, G. (2005). Ultrasonographic and gross pathological findings in the mammary glands of weaned sows having suffered recidiving mastitis metritis agalactia. *Reproduction in Domestic Animals* 40(6):544-547. DOI: 10.1111/j.1439-0531.2005.00629.x.

Balzani, A., Cordell, H. J. and Edwards, S. A. (2016). Relationship of sow udder morphology with piglet suckling behavior and teat access. *Theriogenology* 86(8):1913-1920. DOI: 10.1016/j.theriogenology.2016.06.007.

Barnett, J. L., Hemsworth, P. H., Cronin, G. M., Jongman, E. C. and Hutson, G. D. (2001). A review of the welfare issues for sows and piglets in relation to housing. *Australian Journal of Agricultural Research* 52(1):1-28.

Baumgartner, J. (2020). Pig industry in CH, CZ, DE, DK, NL, NO, SE, UK, AT and EU [Internet]. Available at: https://www.vetmeduni.ac.at/fileadmin/v/tierhaltung/I-.

Baxter, E. M., Rutherford, K. M. D., D'Eath, R. B., Arnott, G., Turner, S. P., Sandøe, P., Moustsen, V., Thorup, F., Edwards, S. A. and Lawrence, A. B. (2013). The welfare implications of large litter size in the domestic pig II: Management factors. *Animal Welfare* 22(2):219-238.

Beaulieu, A. D., Aalhus, J. L., Williams, N. H. and Patience, J. F. (2010). Impact of piglet birth weight, birth order, and litter size on subsequent growth performance, carcass quality, muscle composition, and eating quality of pork. *Journal of Animal Science* 88(8):2767-2778.

Becker, H. A., Kurtz, R. and Mickwitz, G. (1985). Chronische Harnwegsinfektionen beim Schwein, Diagnose und Therapie. *Praktische Tierarzt* 12:1006-1011.

Bellino, C., Gianella, P., Grattarola, C., Miniscalco, B., Tursi, M., Dondo, A., D'Angelo, A. and Cagnasso, A. (2013). Urinary tract infections in sows in Italy: Accuracy of urinalysis and urine culture against histological findings. *Veterinary Record* 172(7):183.

Benthem de Grave, X., Saltzmann, J., Laurain, J., Rodriguez, M. A., Molist, F., Dänicke, S. and Santos, R. R. (2021). Transmission of zearalenone, deoxynivalenol, and their derivatives from sows to piglets during lactation. *Toxins* 13(1):37.

Berner, H. (1981). Untersuchungen zum Vorkommen von Harnwegsinfektionen beim Schwein, 1. Mitteilung: Harnwegsinfektionen bei Muttersauen in Ferkelerzeugerbetrieben. *Tierärztl. Umsch.* 36:162-171.

Berthiller, F., Sulyok, M., Krska, R. and Schuhmacher, R. (2007). Chromatographic methods for the simultaneous determination of mycotoxins and their conjugates in cereals. *International Journal of Food Microbiology* 119(1-2):33-37.

Biksi, I., Takacs, N., Vetesi, F., Fodor, L., Szenci, O. and Fenyo, E. (2002). Association between endometritis and urocystitis in culled sows. *Acta Veterinaria Hungarica* 50(4):413-423.

Björkman, S. and Grahofer, A. (2020). Tools and protocols for managing hyperprolific sows at parturition: Optimizingpiglet survival and sows' reproductive health. In: *Animal Reproduction in Veterinary Medicine*. London: IntechOpen. DOI: 10.5772/intechopen.91337.

Björkman, S., Oliviero, C., Kauffold, J., Soede, N. M. and Peltoniemi, O. A. T. (2018a). Prolonged parturition and impaired placenta expulsion increase the risk of postpartum metritis and delay uterine involution in sows. *Theriogenology* 106:87-92.

Björkman, S., Grahofer, A., Han, T., Oliviero, C. and Peltoniemi, O. A. T. (2018b). Severe udder edema as a cause of reduced colostrum quality and milk production in sows – A case report. In: *10th European Symposium of Porcine Health Management, Barcelona, Spain* (vols. 9-11), pp. 110-111.

Björkman, S., Oliviero, C., Rajala-Schultz, P. J., Soede, N. M. and Peltoniemi, O. A. T. (2017a). The effect of litter size, parity and farrowing duration on placenta expulsion and retention in sows. *Theriogenology* 92:36-44.

Björkman, S., Oliviero, C., Hasan, S. M. K. and Peltoniemi, O. A. T. (2017b). Poster presentations. *Reproduction in Domestic Animals* 52:66-146. DOI: 10.1111/rda.13026.

Boers, A., Taylor, D. J. and Martineau, G.-P. (2019). Miscellaneous bacterial infections. In: Zimmerman, J. J., Karriker, L. A., Ramirez, A., Schwartz, K. J., Stevenson, G. W. and Zhang, J. (Eds) *Diseases of Swine* (11th edn.). Glasgow, Great Britain: Wiley Blackwell (vol. 2019), pp. 981-1001.

Bokma, S., Kersjes, G. J., Unshelm, J., Putten, G. V. and Zeeb, K. (1984). The introduction of pregnant sows in an established group. In: Unshelm, J. (Ed.) *Proceedings of the*

International Congress on Applied Ethology in Farm Animals, Skara, Sweden, pp. 166–169.

Broes, A., Taylor, D. J. and Martineau, G. P. (2019). Miscellaneous bacterial infections. *Diseases of Swine,* pp. 981–1001.

Brown, T. T., Jr., Paul, P. S. and Mengeling, W. L. (1980). Response of conventionally raised weanling pigs to experimental infection with a virulent strain of porcine parvovirus. *American Journal of Veterinary Research* 41(8):1221–1224.

Cabrera, R., Lin, X., Ashwell, M., Moeser, A. and Odle, J. (2013). Early postnatal kinetics of colostral immunoglobulin G absorption in fed and fasted piglets and developmental expression of the intestinal immunoglobulin G receptor. *Journal of Animal Science* 91(1):211–218.

Cabrera, R. A., Lin, X., Campbell, J. M., Moeser, A. J. and Odle, J. (2012). Influence of birth order, birth weight, colostrum and serum immunoglobulin G on neonatal piglet survival. *Journal of Animal Science and Biotechnology* 3(1):42.

Camenisch, U., Lu, Z. H., Vaughan, L., Corboz, L., Zimmermann, D. R., Wittenbrink, M. M., Pospischil, A. and Sydler, T. (2004). Diagnostic investigation into the role of Chlamydiae in cases of increased rates of return to oestrus in pigs. *Veterinary Record* 155(19):593–596. DOI: 10.1136/vr.155.19.593.

Castrén, H., Algers, B., de Passille, A.-M., Rushen, J. and Uvnas-Moberg, K. (1993). Preparturient variation in progesterone, prolactin, oxytocin and somatostatin in relation to nest-building in sows. *Applied Animal Behaviour Science* 38(2):91–102.

Castrén, H., Algers, B., De Passillé, A.-M., Rushen, J. and Uvnäs-Moberg, K. (1994). Nest-building in sows in relation to hormone release. *Applied Animal Behaviour Science* 40(1):74–75.

Chae, C. (2005). A review of porcine circovirus 2-associated syndromes and diseases. *Veterinary Journal* 169(3):326–336.

Coleman, D. N., Lopreiato, V., Alharthi, A. and Loor, J. J. (2020). Amino acids and the regulation of oxidative stress and immune function in dairy cattle. *Journal of Animal Science* 98(Suppl 1):S175–S193. DOI:10.1093/jas/skaa138.

Corzo, C. A., Mondaca, E., Wayne, S., Torremorell, M., Dee, S., Davies, P. and Morrison, R. B. (2010). Control and elimination of porcine reproductive and respiratory syndrome virus. *Virus Research* 154(1–2):185–192. DOI: 10.1016/j.virusres.2010.08.016.

Costermans, N. G. J., Teerds, K. J., Middelkoop, A., Roelen, B. A. J., Schoevers, E. J., van Tol, H. T. A., Laurenssen, B., Koopmanschap, R. E., Zhao, Y., Blokland, M., van Tricht, F., Zak, L., Keijer, J., Kemp, B. and Soede, N. M. (2020). Consequences of negative energy balance on follicular development and oocyte quality in primiparous sows. *Biology of Reproduction* 102(2):388–398. DOI: 10.1093/biolre/ioz175.

D'Allaire, S., Stein, T. E. and Leman, A. D. (1987). Culling patterns in selected Minnesota swine breeding herds. *Canadian Journal of Veterinary Research* 51(4):506–512.

Damm, B. I., Lisborg, L., Vestergaard, K. S. and Vanicek, J. (2003). Nest-building behavioural disturbances and heart rate in farrowing sows kept in crates and Schmid pens. *Livestock Production Science* 80(3):175–187.

Damm, B. I., Vestergaard, K. S., Schrøder-Petersen, D. L. and Ladewig, J. (2000). The effects of branches on prepartum nest-building in gilts with access to straw. *Applied Animal Behaviour Science* 69(2):113–124.

Dänicke, S., Brüssow, K. P., Goyarts, T., Valenta, H., Ueberschär, K. H. and Tiemann, U. (2007). On the transfer of the Fusarium toxins deoxynivalenol (DON) and zearalenone

(ZON) from the sow to the full-term piglet during the last third of gestation. *Food and Chemical Toxicology* 45(9):1565-1574.

Dänicke, S., Valenta, H., Döll, S., Ganter, M. and Flachowsky, G. (2004). On the effectiveness of a detoxifying agent in preventing fusario-toxicosis in fattening pigs. *Animal Feed Science and Technology* 114(1-4):141-157.

Declerck, I., Dewulf, J., Piepers, S., Decaluwé, R. and Maes, D. (2015). Sow and litter factors influencing colostrum yield and nutritional composition. *Journal of Animal Science* 93(3):1309-1317. DOI: 10.2527/jas.2014-8282.

Declerck, I., Dewulf, J., Sarrazin, S. and Maes, D. (2016). Long-term effects of colostrum intake in piglet mortality and performance. *Journal of Animal Science* 94(4):1633-1643.

Declerck, I., Sarrazin, S., Dewulf, J. and Maes, D. (2017). Sow and piglet factors determining variation of colostrum intake between and within litters. *Animal* 11(8):1336-1343.

Dee, S. A. (1992). Porcine urogenital disease. The Veterinary Clinics of North America. *Food Animal Practice* 8(3):641-660. DOI: 10.1016/s0749-0720(15)30709-x.

de Jong, E., Appeltant, R., Cools, A., Beek, J., Boyen, F., Chiers, K. and Maes, D. (2014). Slaughterhouse examination of culled sows in commercial pig herds. *Livestock Science* 167:362-369. DOI: 10.1016/j.livsci.2014.07.001.

De Winter, P. J. J., Verdonck, M., de Kruif, A., de Vriese, L. A. and Haesebrouck, F. (1992). Endometritis and vaginal discharge in the sow. *Animal Reproduction Science* 28(1-4):51-58.

De Winter, P. J. J., Verdonck, M., De Kruif, A., Devriese, L. A. and Haesebrouck, F. (1995). Bacterial endometritis and vaginal discharge in the sow: Prevalence of different bacterial species and experimental reproduction of the syndrome. *Animal Reproduction Science* 37(3-4):325-335.

Dial, G. D. and MacLachlan, N. J. (1988). Urogenital infections of swine. Part I. Clinical manifestations and pathogenesis. *Compendium on Continuing Education for the Practicing Veterinarian* 10:63-68.

Drolet, R. (2019). Urinary system. In: Straw, B. E., Zimmerman, J. J., D'Allaire, S. and Taylor, D. J. (Eds) *Diseases of Swine* (11th edn.). Iowa: Blackwell Publishing.

Eckburg, P. B., Bik, E. M., Bernstein, C. N., Purdom, E., Dethlefsen, L., Sargent, M., Gill, S. R., Nelson, K. E. and Relman, D. A. (2005). Diversity of the human intestinal microbial flora. *Science* 308(5728):1635-1638. DOI: 10.1126/science.1110591.

Edwards, S. A., & Baxter, E. M. (2015). Piglet mortality: causes and prevention. In *The Gestating and Lactating Sow*. Wageningen Academic Publishers, Wageningen, The Netherlands. pp. 649-653.

Einarsson, S., Sjunneson, Y., Hulten, F., Eliasson-Selling, L., Dalin, A. M., Lundeheim, N. and Magnusson, U. A. (2014). 25 years experience of group-housed sows-reproduction in animal welfare-friendly systems. *Acta Veterinaria Scandinavica* 56:37.

Eissen, J. J., Kanis, E. and Kemp, B. (2000). Sow factors affecting voluntary feed intake during lactation. *Livestock Production Science* 64(2-3):147-165.

Eliasson-Selling, L., Hofmo, P. O. and Narum, M. (2000). The art of oestrous-control in loose housed sows. In: Cargill, C. and McOrist, S. (Eds) *Proceedings of the International Pig Veterinary Society Congres*, 17-21 September 2000. Melbourne. Causal Productions Pty, Limited (vol. 2000), p. 395.

Ellendorff, F., Taverne, M., Elsaesser, F., Forsling, M., Parvizi, N., Naaktgeboren, C. and Smidt, D. (1979). Endocrinology of parturition in the pig. *Animal Reproduction Science* 2(1-3):323-334.

Ellis, W. A. (1990). Leptospirosis. In: *OIE Manual of Recommended Diagnostic Techniques and Requirements for Biological Products for List A and B Diseases*, Paris, 27:1-11, sec. 7.

Ellis, W. A. (2006). Leptospirosis. In: Straw, B. , Zimmermann, J. J. , D'Allaire, S. and Taylor, D. J. (Eds) *Diseases of Swine* (9th edn.). London: Blackwell Publishing, pp. 691-700.

Ensley, S. M. and Radke, S. L. (2019). Mycotoxins in grains and feeds. In: Zimmerman, J. J., Karriker, L. A., Ramirez, A., Schwartz, K. J., Stevenson, G. W. and Zhang, J. (Eds) *Diseases of Swine* (11th edn.). Glasgow, Great Britain: Wiley Blackwell, pp. 1055-1071.

Eskola, M., Kos, G., Elliott, C. T., Hajšlová, J., Mayar, S. and Krska, R. (2020). Worldwide contamination of food-crops with mycotoxins: Validity of the widely cited 'FAO estimate'of 25%. *Critical Reviews in Food Science and Nutrition* 60(16):2773-2789.

Farmer, C. (2016). Altering prolactin concentrations in sows. *Domestic Animal Endocrinology* 56(Suppl):S155-S164. DOI: 10.1016/j.domaniend.2015.11.005.

Farmer, C., Maes, D. and Peltoniemi, O. A. T. (2019). The mammary system. In: Chwartz, K., Zimmerman, J., Karriker, L., Ramirez, A., Stevenson, G. and Zhang, J. (Eds) *Diseases of Swine* (11th edn.). London: Wiley Blackwell, pp. 313-338.

Farmer, C. and Quesnel, H. (2009). Nutritional, hormonal, and environmental effects on colostrum in sows. *Journal of Animal Science* 87(13)(Suppl):56-64. Available at: http://jas.fass.org/cgi/content/full/87/13_suppl/56.

Ferrari, C. V., Sbardella, P. E., Bernardi, M. L., Coutinho, M. L., Vaz, I. S., Wentz, I. and Bortolozzo, F. P. (2014). Effect of birth weight and colostrum intake on mortality and performance of piglets after cross-fostering in sows of different parities. *Preventive Veterinary Medicine* 114(3-4):259-266.

Feyera, T. and Theil, P. K. (2017). Energy and lysine requirements and balances of sows during transition and lactation: A factorial approach. *Livestock Science* 201:50-57.

Gerjets, I. and Kemper, N. (2009). Coliform mastitis in sows: A review. *Journal of Swine Health and Production* 17(2):97-105.

Gilbert, C. L., Goode, J. A. and MacGrath, T. J. (1994). Pulsatile secrection of oxytocin during parturition in the pig: Temporal relationship with foetal expulsion. *Journalof Physiology* 475(1):129-137.

Gimsa, U., Tuchscherer, M. and Kanitz, E. (2018). Psychosocial stress and immunity-what can we learn From pig studies? *Frontiers in Behavioral Neuroscience* 12:64. DOI: 10.3389/fnbeh.2018.00064.

Glock, X. T. P. and Bilkei, G. (2005). The effect of postparturient urogenital diseases on the lifetime reproductive performance of sows. *Canadian Veterinary Journal* 46(12):1103-1107.

Gmeiner, K. (2007). *Ultrasonographische Charakterisierung der gesunden und kranken Harnblase bei der Sau*. Doctoral Dissertation, University of Leipzig.

Goyarts, T., Dänicke, S., Brüssow, K. P., Valenta, H., Ueberschär, K. H. and Tiemann, U. (2007). On the transfer of the Fusarium toxins deoxynivalenol (DON) and zearalenone (ZON) from sows to their fetuses during days 35-70 of gestation. *Toxicology Letters* 171(1-2):38-49.

Grahofer, A., Bill, R. and Nathues, H. (2017). Vaginal discharge following artificial insemination of sows in a multisite sow pool system. In: *Proceedings of the 21nd Annual ESDAR Conference*.

Grahofer, A., Björkman, S. and Peltoniemi, O. (2020). Diagnosis of endometritis and cystitis in sows: Use of biomarkers. *Journal of Animal Science* 98(Suppl 1):S107-S116.

Grahofer, A., Meile, A. and Nathues, H. (2019). Detection and evaluation of puerperal disorders in sows after farrowing. In: *Proceedings of theFirst Symposium of the European College of Animal Reproduction*, Vienna, Austria.

Grahofer, A., Sipos, S., Fischer, L., Entenfellner, F. and Sipos, W. (2014). Relationship between bacteriological and chemic-analytical urinalysis from sows with reproductive disorders. In: *Proceedings of the Sixth European Symposium of Porcine Health Management* (vol. 126).

Gustafsson, M., Jensen, P., de Jonge, F. H., Illman, G. and Špinka, M. (1999). Maternal behaviour of domestic sows and crosses between domestic sows and wild boar. *Applied Animal Behaviour Science* 65(1):29–42.

Han, T., Björkman, S., Soede, N. M., Oliviero, C. and Peltoniemi, O. (2019). Effect of IGF-1 level at weaning on subsequent luteal developement and progesterone production in primiparous sows. In: *Abstract Book 11th European Symposium of Porcine Health Management; 22-24 May. Utrecht*. The Netherlands: ESPHM, p. 82.

Han, T., Björkman, S., Soede, N. M., Oliviero, C. and Peltoniemi, O. A. T. (2020). IGF-1 concentration patterns and their relationship with follicle development after weaning in young sows fed different pre-mating diets. *Animal* 14(7):1493-1501. DOI: 10.1017/S1751731120000063.

Han, T., Björkman, S., Soede, N. M., Oliviero, C. and Peltoniemi, O. A. T. (2021). IGF-1 concentrations after weaning in young sows fed different pre-mating diets are positively associated with piglet mean birth weight at subsequent farrowing. *Animal* 15(1):100029. DOI: 10.1016/j.animal.2020.100029.

Hartsock, T. G. and Barczewski, R. A. (1997). Prepartum behaviour in swine: Effects of pen size. *Journal of Animal Science* 75(11):2899-2904.

Hasan, S., Junnikkala, S., Peltoniemi, O., Paulin, L., Lyyski, A., Vuorenmaa, J. and Oliviero, C. (2018a). Dietary supplementation with yeast hydrolysate in pregnancy influences colostrum yield and gut microbiota of sows and piglets after birth. *PLoS ONE* 13(5):e0197586. DOI: 10.1371/journal.pone.0197586.

Hasan, S., Saha, S., Junnikkala, S., Orro, T., Peltoniemi, O. and Oliviero, C. (2018b). Late gestation diet supplementation of resin acid-enriched composition increases sow colostrum immunoglobulin G content, piglet colostrum intake and improve sow gut microbiota. *Animal* 27: 1-8. DOI: 10.1017/S1751731118003518.

Hermansson, I., Einarsson, S., Larsson, K. and Backstrom, L. (1978). On the agalactia post partum in the sow. A clinical study. *Nordisk Veterinaermedicin* 30(11):465-473.

Hopper, S. A., White, M. E. and Twiddy, N. (1992). An outbreak of blue-eared pig disease (porcine reproductive and respiratory syndrome) in four pig herds in Great Britain. *Veterinary Record* 131(7):140-144. DOI: 10.1136/vr.131.7.140.

Hoving, L. L., Soede, N. M., Feitsma, H. and Kemp, B. (2012). Lactation weight loss in primiparous sows: Consequences for embryo survival and progesterone and relations with metabolic profiles. *Reproduction in Domestic Animals* 47(6):1009-1016.

Hoving, L. L., Soede, N. M., Graat, E. A. M., Feitsma, H. and Kemp, B. (2011). Reproductive performance of second parity sows: Relations with subsequent reproduction. *Livestock Science* 140(1-3):124-130.

Hoy, S. (2006). The impact of puerperal diseases in sows on their fertility and health up to next farrowing. *Animal Science* 82(5):701-704. DOI: 10.1079/ASC200670.

Hoy, S., Lutter, C., Wähner, M. and Puppe, B. (1994). The effect of birth weight on the early postnatal vitality of piglets. *DTW. Deutsche Tierarztliche Wochenschrift* 101(10):393-396.

Hultén, F., Persson, A., Eliasson-Selling, L., Heldmer, E., Lindberg, M., Sjögren, U., Kugelberg, C. and Ehlorsson, C. J. (2004). Evaluation of environmental and management-related risk factors associated with chronic mastitis in sows. *American Journal of Veterinary Research* 65(10):1398–1403. DOI: 10.2460/ajvr.2004.65.1398.

Islas-Fabila, P., Mota-Rojas, D., Martínez-Burnes, J., Mora-Medina, P., González-Lozano, M., Roldan-Santiago, P., Greenwell-Beare, V., González-Hernández, M., Vega-Manríquez, X. and Orozco-Gregorio, H. (2018). Physiological and metabolic responses in newborn piglets associated with the birth order. *Animal Reproduction Science* 197:247–256.

Jackson, P. G. G. (1995). *Handbook of Veterinary Obstetrics* (vol. 1995, 2nd edn.). London: Saunders, pp. 221–222.

Jarvis, S., Lawrence, A. B., McLean, K. A., Deans, L. A., Chirnside, J. and Calvert, S. K. (1997). The effect of environment on behavioural activity, ACTH, beta-endorphin and cortisol in prefarrowing gilts. *Animal Science* 65(3):465–472.

Ji, Y. J., Li, H., Xie, P. F., Li, Z. H., Li, H. W., Yin, Y. L., Blachier, F. and Kong, X. F. (2019). Stages of pregnancy and weaning influence the gut microbiota diversity and function in sows. *Journal of Applied Microbiology* 127(3):867–879. DOI: 10.1111/jam.14344.

Jost, T., Lacroix, C., Braegger, C. and Chassard, C. (2014). Stability of the maternal gut microbiota during late pregnancy and early lactation. *Current Microbiology* 68(4):419–427.

Kaeoket, K., Persson, E. and Dalin, A. M. (2001). The sow endometrium at different stages of the oestrus cycle: Studies on morphological changes and infiltration by cells of the immune system. *Animal Reproduction Science* 65(1–2):95–114.

Kaiser, M., Jacobson, M., Andersen, P. H., Bækbo, P., Cerón, J. J., Dahl, J., Escribano, D. and Jacobsen, S. (2018a). Inflammatory markers before and after farrowing in healthy sows and in sows affected with postpartum dysgalactia syndrome. *BMC Veterinary Research* 14(1):83. DOI: 10.1186/s12917-018-1382-7.

Kaiser, M., Jacobsen, S., Andersen, P. H., Bækbo, P., Cerón, J. J., Dahl, J., Escribano, D., Theil, P. K. and Jacobson, M. (2018b). Hormonal and metabolic indicators before and after farrowing in sows affected with postpartum dysgalactia syndrome. *BMC Veterinary Research* 14(1):334. DOI: 10.1186/s12917-018-1649-z.

Kaiser, M., Jacobson, M., Bækbo, P., Dahl, J., Jacobsen, S., Guo, Y. Z., Larsen, T. and Andersen, P. H. (2020). Lack of evidence of mastitis as a causal factor for postpartum dysgalactia syndrome in sows. *Translational Animal Science* 4(1):250–263.

Kanora, A. and Maes, D. (2010). The role of mycotoxins in pig reproduction: A review. *Veterinarni Medicina* 54(12):565–576.

Kapell, D. N. R. G., Ashworth, C. J., Knap, P. W. and Roehe, R. (2011). Genetic parameters for piglet survival, litter size and birth weight or its variation within litter in sire and dam lines using Bayesian analysis. *Livestock Science* 135(2–3):215–224.

Kauffold, J. (2008). Nichtpuerperale Uterusentzündungen beim Schwein. *Tierärztliche Praxis Ausgabe G* 36(3):189–198.

Kauffold, J. and Althouse, G. C. (2007). An update on the use of B-mode ultrasonography in female pig reproduction. *Theriogenology* 67(5):901–911.

Kauffold, J., Gmeiner, K., Sobiraj, A., Richter, A., Failing, K. and Wendt, M. (2010). Ultrasonographic characterization of the urinary bladder in sows with and without urinary tract infection. *Veterinary Journal* 183(1):103–108.

Kemp, B., Da Silva, C. L. A. and Soede, N. M. (2018). Recent advances in pig reproduction: Focus on impact of genetic selection for female fertility. *Reproduction in Domestic Animals* 53(Suppl 2):28–36.

Kemp, B. and Soede, N. M. (2012). Reproductive issues in welfare-friendly housing systems in Pig husbandry: A review. *Reproduction in Domestic Animals* 47(Suppl 5):51–57.

Kemper, N. (2020). Update on postpartum dysgalactia syndrome in sows. *Journal of Animal Science* 98(Suppl 1):S117–S125. DOI: 10.1093/jas/skaa135.

Kielland, C., Rootwelt, V., Reksen, O. and Framstad, T. (2015). The association between immunoglobulin G in sow colostrum and piglet plasma. *Journal of Animal Science* 93(9):4453–4462.

Kim, H. B., Borewicz, K., White, B. A. and Singer, R. S. (2011). Longitudinal investigation of the age-related bacterial diversity in the feces of commercial pigs. *Veterinary Microbiology* 153(1–2):124–133.

Kindahl, H., Alonso, R., Cort, N. and Einarsson, S. (1982). Release of prostaglandin F2α during parturition in the sow. *Zentralblatt Fur Veterinarmedizin. Reihe A* 29(7):504–510.

Klobasa, F., Werhahn, E. and Butler, J. E. (1987). Composition of sow milk during lactation. *Journalof Animal Science* 64(5):1458–1466.

Knap, P. W. and Doeschl Wilson, A. (2021). Control and elimination of porcine reproductive and respiratory syndrome virus. *Virus Research* 154(1–2):185–192. DOI: 10.1016/j.virusres.2010.08.016.

Koren, O., Goodrich, J. K., Cullender, T. C., Spor, A., Laitinen, K., Bäckhed, H. K., Gonzalez, A., Werner, J. J., Angenent, L. T., Knight, R., Bäckhed, F., Isolauri, E., Salminen, S. and Ley, R. E. (2012). Host remodeling of the gut microbiome and metabolic changes during pregnancy. *Cell* 150(3):470–480. DOI: 10.1016/j.cell.2012.07.008.

Kraft, W., Dürr, U. M., Fürll, M., Bostedt, H. and Heinritzi, K. (2005). Harnapparat. In: Kraft, W. and Dürr, U. M. (Eds) *Klinische Labordiagnostik in der Tiermedizin* (vol. 6). Aufl. Stuttgart: Verlag Schattauer. S, pp. 186–217.

Labroue, F., Caugant, A., Ligonesche, B. and Gaudré, D. (2001). Étude de l'évolution des tétines dápparence douteuse chez la cochette au cours de sa carrier. *Journées de la Recherche Porcine en France* 33:145–150.

Lager, K. M. and Halbur, P. G. (1996). Gross and microscopic lesions in porcine fetuses infectedwith porcine reproductive and respiratory syndrome virus. *Journal of Veterinary Diagnostic Investigation* 8(3):275–282.

Larsen, I. and Thorup, F. (2006). The diagnosis of MMA. *The International Pig Veterinary Society* 52:256.

Lawlor, P. G. and Lynch, P. B. (2001). Mycotoxins in pig feeds-1: Source of toxins, prevention and management of mycotoxicosis. *Irish Veterinary Journal* 54(3):117–120.

Le Dividich, J., Charneca, R. and Thomas, F. (2017). Relationship between birth order, birth weight, colostrum intake, acquisition of passive immunity and pre-weaning mortality of piglets. *Spanish Journal of Agricultural Research* 15(2):e0603. DOI: 10.5424/sjar/2017152-9921.

Lessard, M., Blais, M., Beaudoin, F., Deschene, K., Lo Verso, L. L., Bissonnette, N., Lauzon, K. and Guay, F. (2018). Piglet weight gain during the first two weeks of lactation influences the immune system development. *Veterinary Immunology and Immunopathology* 206:25–34. DOI: 10.1016/j.vetimm.2018.11.005.

Li, H., Liu, Z., Lyu, H., Gu, X., Song, Z., He, X. and Fan, Z. (2020). Effects of dietary insulin during late gestation on sow physiology, farrowing duration and piglet performance. *Animal Reproduction Science* 219:106531.

Liebhold, M., Wendt, M., Kaup, F. J. and Drommer, W. (1995). Clinical and light and electron microscopical findings in sows with cystitis. *Veterinary Record* 137(6):141-144.

Lindsay, D. S., Dubey, J. P. and Santin-Duran, M. (2019). Coccidia and other protozoa. In: Zimmerman, J. J., Karriker, L. A., Ramirez, A., Schwartz, K. J., Stevenson, G. W. and Zhang, J. (Eds) *Diseases of Swine* (11th edn.). Glasgow, Great Britain: Wiley Blackwell (vol. 2019), pp. 1015-1027.

Litvak, Y., Byndloss, M. X., Tsolis, R. M. and Bäumler, A. J. (2017). Dysbiotic proteobacteria expansion: A microbial signature of epithelial dysfunction. *Current Opinion in Microbiology* 39:1-6. DOI: 10.1016/j.mib.2017.07.003.

Liu, P., Piao, X. S., Kim, S. W., Wang, L., Shen, Y. B., Lee, H. S. and Li, S. Y. (2008). Effects of chito-oligosaccharide supplementation on the growth performance, nutrient digestibility, intestinal morphology, and fecal shedding of and in weaning pigs. *Journal of Animal Science* 86(10):2609-2618.

Love, R. J., Evans, G. and Klupiec, C. (1993). Seasonal effects on fertility in gilts and sows. *Journal of Reproduction and Fertility Supplement* 48:191-206.

Luescher, U. A., Friendship, R. M. and McKeown, D. B. (1990). Evaluation of methods to reduce fighting among regrouped gilts. *Canadian Journal of Animal Science* 70(2):363-370.

Madec, F. and Leon, E. (1992). Farrowing disorders in the sow: A field study. *Zentralblatt Fur Veterinarmedizin. Reihe A* 39(6):433-444.

Maes, D., Papadopoulos, G., Cools, A. and Janssens, G. P. (2010). Postpartum dysgalactia in sows: Pathophysiology and risk factors. *Tierärztliche Praxis Ausgabe G: Großtiere/ Nutztiere* 38(1):15-20.

Maes, D., Pluym, L. and Peltoniemi, O. (2016). Impact of group housing of pregnant sows on health. *Porcine Health Management* 2(17):17. DOI:10.1186/s40813-016-0032-3.

Maldonado, J., Segales, J., Martınez-Puig, D., Calsamiglia, M., Riera, P., Domingo, M. and Artigas, C. (2005). Identification of viral pathogens in aborted fetuses and stillborn piglets from cases of swine reproductive failure in Spain. *Veterinary Journal* 169(3):454-456.

Manjarin, R., Montano, Y. A., Kirkwood, R. N., Bennet, D. C. and Petrovski, K. R. (2018). Effect of piglet separation from dam at birth on colostrum uptake. *Canadian Journal of Veterinary Research* 82(3):239-242.

Matheson, S. M., Walling, G. A. and Edwards, S. A. (2018). Genetic selection against intrauterine growth retardation in piglets: A problem at the piglet level with a solution at the sow level. *Genetics, Selection, Evolution* 50(1):46.

Mauget, R. (1982). Seasonality of reproduction in the wild boar. In: Foxcroft, J. and Cole, R. (Eds) *Control of Pig Reproduction*. London: Butterworths, pp. 509-526.

Meile, A., Nathues, H. and Grahofer, A. (2019). Association between uterine involution in sows and reproductive performance in their next gestation. In: *Proceedings of the 17th International Conference on Production Diseases in Farm Animals*.

Menard, J., Batista, L. and D'Allaire, S. D. (2007). Gilt exposure to homologous Porcine Reproductive and Respiratory Syndrome (PRRS) virus strain during acclimatization as a tool for PRRS control. PMWS, PRRS and Swine influenza associated diseases. In: *5th international symposium on emerging and re-emerging pig diseases*, Krakow, Poland (vol. 1), pp. 24th-27th.

Menegat, M. B., DeRouchey, J. M., Woodworth, J. C., Dritz, S. S., Tokach, M. D. and Goodband, R. D. (2019). Effects of Bacillus subtilis C-3102 on sow and progeny performance, fecal consistency, and fecal microbes during gestation, lactation, and nursery periods. *Journal of Animal Science* 97(9):3920-3937. DOI: 10.1093/jas/skz236.

Mengeling, W. L., Lager, K. M. and Vorwald, A. C. (2000). The effect of porcine parvovirus and porcine reproductive and respiratory syndrome virus on porcine reproductive performance. *Animal Reproduction Science* 60-61:199-210.

Meredith, M. J. (1991). Non specific bacterial infections of the genital tract in female pigs. *Pig Veternery Journal* 27:110-121.

Mettenleiter, T. C., Ehlers, B., Müller, T., Yoon, K. J. and Teifke, J. P. (2019). Herpesviruses. In: Zimmerman, J. J., Karriker, L. A., Ramirez, A., Schwartz, K. J., Stevenson, G. W. and Zhang, J. (Eds) *Diseases of Swine* (11th edn.). Glasgow, Great Britain: Wiley Blackwell (vol. 2019), pp. 549-575.

Meulen, J., Helmond, F. A., Oudenaarden, C. P. J. and Van der Meulen, J. (1988). Effect of flushing of blastocysts on days 10-13 on the life-span of the corpora lutea in the pig. *Journal of Reproduction and Fertility* 84(1):157-162.

Moore, A. S., Gonyou, H. W. and Ghent, A. W. (1993). Integration of newly introduced and resident sows following grouping. *Applied Animal Behaviour Science* 38(3-4):257-267.

Moreira, L. P., Menegat, M. B., Barros, G. P., Bernardi, M. L., Wentz, I. and Bortolozzo, F. P. (2017). Effects of colostrum, and protein and energy supplementation on survival and performance of low-birth-weight piglets. *Livestock Science* 202:188-193.

Morgan, X. C., Tickle, T. L., Sokol, H., Gevers, D., Devaney, K. L., Ward, D. V., Reyes, J. A., Shah, S. A., LeLeiko, N., Snapper, S. B., Bousvaros, A., Korzenik, J., Sands, B. E., Xavier, R. J. and Huttenhower, C. (2012). Dysfunction of the intestinal microbiome in inflammatory bowel disease and treatment. *Genome Biology* 13(9):R79. DOI: 10.1186/gb-2012-13-9-r79.

Muirhead, M. R. (1986). Epidemiology and control of vaginal discharges in the sow after service. *The Veterinary Record* 119(10):233-235.

Munsterhjelm, C., Valros, A., Heinonen, M., Hälli, O. and Peltoniemi, O. A. T. (2008). Housing during early pregnancy affects fertility and behaviour of sows. *Reproduction in Domestic Animals* 43(5):584-591.

Nachreiner, R. F. and Ginther, O. J. (1974). Induction of agalactia by administration of endotoxin (Escherichia coli) in swine. *American Journal of Veterinary Research* 35(5):619-622.

Nagy, G. (1993). Comparative pathogenicity study of *Leptospira interrogans* serovar Pomona strains. *Acta Veterinaria Hungarica* 41(3-4):315-324.

Nash, W. A. (1990). Porcine parvovirus survey. *Veterinary Record* 126(7):175-176.

Neumann, E. J. and Hall, W. F. (2019). Disease control, prevention, and elimination. In: Zimmerman, J. J., Karriker, L. A., Ramirez, A., Schwartz, K. J., Stevenson, G. W. and Zhang, J. (Eds) *Diseases of Swine* (11th edn.). Glasgow, Great Britain: Wiley Blackwell (vol. 2019), pp. 123-157.

Nguyen, K., Cassar, G., Friendship, R. M. and Hodgins, D. (2013). An investigation of the impacts of induced parturition, birth weight, birth order, litter size, and sow parity on piglet serum concentrations of immunoglobulin G. *Journal of Swine Health and Production* 21:139-143.

OIE. (2008). Leptospirosis. In: *Manual of Diagnostic Tests and Vaccines for Terrestrial Animals* (6th edn.). Paris: Office International Des Epizooties (vol. 1), pp. 251-264.

Oliviero, C. (2010). Successful farrowing. PhD thesis, University of Helsinki. Available at: https://helda.helsinki.fi/bitstream/handle/10138/18971/successf.pdf?sequence=1 &isAllowed=y.

Oliviero, C., Pastell, M., Heinonen, M., Heikkonen, J., Valros, A., Ahokas, J., Vainio, O. and Peltoniemi, O. A. T. (2008a). Using movement sensors to detect the onset of farrowing. *Biosystems Engineering* 100(2):281–285.

Oliviero, C., Heinonen, M., Valros, A., Halli, O. and Peltoniemi, O. A. T. (2008b). Effect of the environment on the physiology of the sow during late pregnancy, farrowing and early lactation. *Animal Reproduction Science* 105(3–4):365–377.

Oliviero, C., Heinonen, M., Valros, A. and Peltoniemi, O. A. T. (2010). Environmental and sow-related factors affecting the duration of farrowing. *Animal Reproduction Science* 119(1–2):85–91.

Oliviero, C., Junnikkala, S. and Peltoniemi, O. A. T. (2019). The challenge of large litters on the immune system of the sow and the piglets. *Reproduction in Domestic Animals* 54(Suppl 3):12–21. DOI: 10.1111/rda.13463.

Oliviero, C., Kokkonen, T., Heinonen, M., Sankari, S. and Peltoniemi, O. A. T. (2009). Feeding sows a high-fibre diet around farrowing and early lactation: Impact on intestinal activity, energy balance-related parameters and litter performance. *Research in Veterinary Science* 86(2):314–319.

Oliviero, C., Kothe, S., Heinonen, M., Valros, A. and Peltoniemi, O. A. T. (2013). Prolonged duration of farrowing is associated with subsequent decreased fertility in sows. *Theriogenology* 79(7):1095–1099.

Olsen, S. C., Nol, P. and Samartino, L. (2019). Brucellosis. In: Zimmerman, J. J., Karriker, L. A., Ramirez, A., Schwartz, K. J., Stevenson, G. W. and Zhang, J. (Eds) *Diseases of Swine* (11th edn.). Glasgow, Great Britain: Wiley Blackwell (vol. 2019), pp. 778–791.

Opriessnig, T. and Coutinho, T. A. (2019). Erysipelas. In: Zimmerman, J. J., Karriker, L. A., Ramirez, A., Schwartz, K. J., Stevenson, G. W. and Zhang, J. (Eds) *Diseases of Swine* (11th edn.). Glasgow, Great Britain: Wiley Blackwell (vol. 2019), pp. 635–843.

Oravainen, J., Heinonen, M., Seppä-Lassila, L., Orro, T., Tast, A., Virolainen, J. V. and Peltoniemi, O. A. T. (2006). Factors affecting fertility in loosely housed sows and gilts: Vulvar discharge syndrome, environment and acute-phase proteins. *Reproduction in Domestic Animals* 41(6):549–554.

Oravainen, J., Heinonen, M., Tast, A., Virolainen, J. V. and Peltoniemi, O. A. T. (2005). High porcine parvovirus antibodies in sow herds: Prevalence and associated factors. *Reproduction in Domestic Animals* 40(1):57–61.

Oravainen, J., Heinonen, M., Tast, A., Virolainen, J. V. and Peltoniemi, O. A. T. (2008). Vulvar discharge syndrome in loosely housed Finnish pigs: Prevalence and evaluation of vaginoscopy, bacteriology and cytology. *Reproduction in Domestic Animals* 43(1):42–47. DOI: 10.1111/j.1439-0531.2007.00852.x.

Papadopoulos, G. A., Vanderhaeghe, C., Janssens, G. P., Dewulf, J. and Maes, D. G. (2010). Risk factors associated with postpartum dysgalactia syndrome in sows. *Veterinary Journal* 184(2):167–171. DOI: 10.1016/j.tvjl.2009.01.010.

Peltoniemi, O. A. T., Björkman, S., Oropeza-Moe, M. and Oliviero, C. (2019). Developments of reproductive management and biotechnology in the pig. *Animal Reproduction* 16(3):524–538. DOI: 10.21451/1984-3143-AR2019-0055.

Peltoniemi, O. A. T., Oliviero, C., Yun, J., Grahofer, A. and Björkman, S. (2020). Management practices to optimize the parturition process in the hyperprolific sow. *Journal of Animal Science* 98(Suppl 1):S96–S106.

Peltoniemi, O. A. T., Love, R. J., Heinonen, M., Tuovinen, V. and Saloniemi, H. (1999). Seasonal and management effects on fertility of the sow: A descriptive study. *Animal Reproduction Science* 55(1):47-61.

Peltoniemi, O. A. T., Björkman, S. and Maes, D. (2016). Reproduction of group-housed sows. *Porcine Health Management* 2:15. DOI: 10.1186/s40813-016-0033-2.

Peltoniemi, O. A. T., Love, R. J., Klupiec, C. and Evans, G. (1997). Effect of feed restriction and season on LH and prolactin secretion, adrenal response, insulin and FFA in group housed pregnant gilts. *Animal Reproduction Science* 49(2-3):179-190.

Peltoniemi, O. A. T., Tast, A. and Love, R. J. (2000). Factors effecting reproduction in the pig:seasonal effects and restricted feeding of the pregnant gilt and sow. *Animal Reproduction Science* 60-61:173-184.

Peltoniemi, O. A. T. and Virolainen, J. V. (2006). Seasonality of reproduction in gilts and sows. *Society of Reproduction and Fertility Supplement* 62:205-218.

Pensart, M. B. and Kluge, J. P. (1989). Pseudorabies virus (Aujeszky's disease). In: Pensart, M. B. (Ed.) *Virus Infections in Porcines*. New York: Elsevier, pp. 39-64.

Petersen, H. H., Nielsen, J. P. and Heegaard, P. M. (2004). Application of acute phase protein measurements in veterinary clinical chemistry. *Veterinary Research* 35(2):163-187. DOI: 10.1051/vetres:2004002.

Quesnel, H., Brossard, L., Valancogne, A. and Quiniou, N. (2008). Influence of some sow characteristics on within-litter variation of piglet birth weight. *Animal* 2(12):1842-1849.

Quesnel, H., Farmer, C. and Devillers, N. (2012). Colostrum intake: Influence on piglet performance and factors of variation. *Livestock Science* 146(2-3):105-114.

Quiniou, N., Dagorn, J. and Gaudré, D. (2002). Variation of piglets' birth weight and consequences on subsequent performance. *Livestock Production Science* 78(1):63-70.

Remience, V., Wavreille, J., Canart, B., Meunier-Salaün, M. C., Prunier, A., Bartiaux-Thill, N., Nicks, B. and Vandenheede, M. (2008). Effects of space allowance on the welfare of dry sows kept in dynamic groups and fed with an electronic sow feeder. *Applied Animal Behaviour Science* 112(3-4):284-296.

Rooke, J. A. and Bland, I. M. (2002). The acquisition of passive immunity in the new-born piglet. *Livestock Production Science* 78(1):13-23.

Rozeboom, D. W. (2015). Conditioning of the gilt for optimal reproductive performance. In: *The Gestating and Lactating Sow*. Editor: Farmer C. Wageningen Academic Publishers, Wageningen, The Netherlands. pp. 649-653.

Rutherford, K. M., Baxter, E. M., D'Eath, R. B., Turner, S. P., Arnott, G., Roehe, R., Ask, B., Sandøe, P., Moustsen, V. A., Thorup, F., Edwards, S. A., Berg, P. and Lawrence, A. B. (2013). The welfare implications of large litter size in the domestic pig I: Biological factors. *Animal Welfare* 22(2):199-218.

Salmon, H., Berri, M., Gerdts, V. and Meurens, F. (2009). Humoral and cellular factors of maternal immunity in swine. *Developmental and Comparative Immunology* 33(3):384-393. DOI: 10.1016/j.dci.2008.07.007.

Saporiti, V., Valls, L., Maldonado, J., Perez, M., Correa-Fiz, F., Segalés, J. and Sibila, M. (2021). Porcine Circovirus 3 detection in aborted fetuses and stillborn piglets from swine reproductive failure cases. *Viruses* 13(2):264. DOI: 10.3390/v13020264.

Sassone-Corsi, M. and Raffatellu, M. (2015). No vacancy: How beneficial microbes cooperate with immunity to provide colonization resistance to pathogens. *Journal of Immunology* 194(9):4081-4087.

Sayyari, A., Uhlig, S., Fæste, C. K., Framstad, T. and Sivertsen, T. (2018). Transfer of deoxynivalenol (DON) through placenta, colostrum and milk from sows to their offspring during late gestation and lactation. *Toxins* 10(12):517.

Schoos, A., Chantziaras, I., Vandenabeele, J., Biebaut, E., Meyer, E., Cools, A., Devreese, M. and Maes, D. (2020). Prophylactic use of meloxicam and paracetamol in peripartal sows suffering From postpartum Dysgalactia syndrome. *Frontiers in Veterinary Science* 7:603719. DOI: 10.3389/fvets.2020.603719.

Segalés, J. (2012). Porcine circovirus type 2 (PCV2) infections: Clinical signs, pathology and laboratory diagnosis. *Virus Research* 164(1–2):10–19.

Shin, N. R., Whon, T. W. and Bae, J. W. (2015). Proteobacteria: Microbial signature of dysbiosis in gut microbiota. *Trends in Biotechnology* 33(9):496–503. DOI: 10.1016/j. tibtech.2015.06.011.

Sipos, W., Grahofer, A., Fischer, L., Entenfellner, F. and Sipos, S. (2014). Keimspektrum des Urogenitaltraktes von Sauen mit Fertilitätsstörungen. *Wien Tierarztl Monat* 101: 9–10, 214–220.

Smit, M. N., Spencer, J. D., Almeida, F. R., Patterson, J. L., Chiarini-Garcia, H., Dyck, M. K. and Foxcroft, G. R. (2013). Consequences of a low litter birth weight phenotype for postnatal lean growth performance and neonatal testicular morphology in the pig. *Animal* 7(10):1681–1689. DOI: 10.1017/S1751731113001249.

Soede, N. M. and Kemp, B. (2015). Best practices in the lactating and weaned sow to optimize reproductive physiology and performance. In: *The Gestating and Lactating Sow*. Editor: Farmer C. Wageningen Academic Publishers, Wageningen, The Netherlands. pp. 649–653.

Spoolder, H. A., Burbidge, J. A., Edwards, S. A., Lawrence, A. B. and Simmins, P. H. (1996). Social recognition in gilts mixed into a dynamic group of 30 sows. In: *Proceedings of the British Society of Animal Science* (vol. 1996). Cambridge: Cambridge University Press, p. 37.

Spoolder, H. A. and Vermeer, H. M. (2015). Gestation group housing of sows. In: *The Gestating and Lactating Sow*. Editor: Farmer C. Wageningen Academic Publishers, Wageningen, The Netherlands. pp. 649–653.

Stiehler, T., Heuwieser, W., Pfuetzner, A. and Burfeind, O. (2015). The course of rectal and vaginal temperature in early postpartum sows. *Journal of Swine Health and Production* 23:72–83.

Stookey, J. M. and Gonyou, H. W. (1998). Recognition in swine: Recognition through familiarity or genetic relatedness? *Applied Animal Behaviour Science* 55(3–4):291–305.

Tan, C., Wei, H., Ao, J., Long, G. and Peng, J. (2016). Inclusion of konjac Flour in the gestation diet changes the gut microbiota, alleviates oxidative stress, and improves insulin sensitivity in sows. *Applied and Environmental Microbiology* 82(19):5899–5909.

Tast, A., Halli, O., Ahlstrom, S., Andersson, H., Love, R. J. and Peltoniemi, O. A. T. (2001). Seasonal alterations in circadian melatonin rhythms of the European wild boar and domestic gilt. *Journal of Pineal Research* 30(1):43–49.

Tast, A., Peltoniemi, O. A. T., Virolainen, J. V. and Love, R. J. (2002). Early disruption of pregnancy as a manifestation of seasonal infertility in pigs. *Animal Reproduction Science* 74(1–2):75–86.

Taverne, M. A. M. and van der Weijden, G. C. (2008). Parturition in domestic animals: Targets for future research. *Reproduction in Domestic Animals* 43(Suppl 5):36–42.

Thaker, M. Y. and Bilkei, G. (2005). Lactation weight loss influences subsequent reproductive performance of sows. *Animal Reproduction Science* 88(3-4):309-318.

Theil, P. K., Lauridsen, C. and Quesnel, H. (2014). Neonatal piglet survival: Impact of sow nutrition around parturition on fetal glycogen deposition and production and composition of colostrum and transient milk. *Animal* 8(7):1021-1030.

Thorsen, C. K., Schild, S. A., Rangstrup-Christensen, L., Bilde, T. and Pedersen, L. J. (2017). The effect of farrowing duration on maternal behavior of hyper-prolific sows in organic outdoor production. *Livestock Science* 204:92-97.

Tolstrup, L. K. (2017). *Cystitis in Sows: Prevalence, Diagnosis and Reproductive Effect.* Doctoral dissertation, University of Copenhagen.

Tsuma, V. T., Einarsson, S., Madej, A., Kindahl, H. and Lundeheim, N. (1996). Effect of food deprivation during early pregnancy on endocrine changes in primiparous sows. *Animal Reproduction Science* 41(3-4):267-278.

Tummaruk, P., Kesdangsakonwut, S., Prapasarakul, N. and Kaeoket, K. (2010). Endometritis in gilts: Reproductive data, bacterial culture, histopathology, and infiltration of immune cells in the endometrium. *Comparative Clinical Pathology* 19(6):575-584.

Vasdal, G. and Andersen, I. L. (2012). A note on teat accessibility and sow parity– Consequences for newborn piglets. *Livestock Science* 146(1):91-94. DOI: 10.1016/j. livsci.2012.02.005.

Völkl, A., Vogler, B., Schollenberger, M. and Karlovsky, P. (2004). Microbial detoxification of mycotoxin deoxynivalenol. *Journal of Basic Microbiology* 44(2):147-156.

Waldmann, K. H. and Wendt, M. (2001). *Lehrbuch der Schweinekrankheiten* (4th edn.). Stuttgart: Parey Verlag, p. 608.

Weiss, E. (2007). Harnorgane. In: Dahme, E. and Weiss, E. (Eds) *Grundriß der speziellen pathologischen Anatomie der Haustiere.* Enke Verlag: Stuttgart (vol. 6). Aufl., S., pp. 197-199.

Wendt, M., Liebhold, M., Kaup, F., Amtsberg, G. and Bollwahn, W. (1990). Corynebacterium-suis-Infektion beim Schwein, 1. Mitteilung: Klinische Diagnose unter besonderer Berücksichtigung von Harnuntersuchung und Zystoskopie. *Tierärztl Prax* 18(4):353-357.

Winkler, I. (1987). *Zytodiagnostik des puerperalen Zervikalfluors beim Schwein.* Hannover.

Xu, C., Peng, C., Zhang, X. and Peng, J. (2020). Inclusion of soluble fiber in the gestation diet changes the gut microbiota, affects plasma propionate and odd-chain fatty acids levels, and improves insulin sensitivity in sows. *International Journal of Molecular Sciences* 21(2):635.

Yun, J., Ollila, A., Valros, A., Larenza Menzies, P., Heinonen, M., Oliviero, C. and Peltoniemi, O. A. T. (2019). Behavioural alterations in piglets after surgical castration: Effects of analgesia and anaesthesia. *Research in Veterinary Science* 125:36-42. DOI: 10.1016/j.rvsc.2019.05.009.

Yun, J., Swan, K. M., Vienola, K., Kim, Y. Y., Oliviero, C., Peltoniemi, O. A. T. and Valros, A. (2014). Farrowing environment has an impact on sow metabolic status and piglet colostrum intake in early lactation. *Livestock Science* 163:120-125. DOI: 10.1016/j. livsci.2014.02.014.

Zimmerman, J. J., Dee, S. A., Holtkamp, D. J., Murtaugh, M. P., Stadejek, T., Stevenson, G. W., Torremorell, M., Yang, H. and Zhang, J. (2019). Porcine reproductive and respiratory syndrome viruses (Porcine arteriviruses). In: Zimmerman, J. J., Karriker, L. A., Ramirez, A., Schwartz, K. J., Stevenson, G. W. and Zhang, J. (Eds) *Diseases of Swine* (11th edn.). Glasgow, Great Britain: Wiley Blackwell (vol. 2019), pp. 685-708.

Chapter 2

Optimising pig welfare in breeding and gestation

Paul H. Hemsworth, University of Melbourne, Australia

1 Introduction

The basic social group of wild boars is typically organised around a nucleus of closely related females and their most recent litters (Dardallion, 1988). These matriarchal groups consist of related subadult (younger than one year) and adult (older than one year) females and their piglets, and groups of up to 30-40 individuals have been reported in non-hunted areas (Leaper et al., 1999; Kramer-Schadt et al., 2007). The groups are spatially restricted to home ranges of about 5-20 km² (Leaper et al., 1999) and although the respective home ranges may overlap, these matriarchal groups do not appear to associate with other families, avoiding contact through a strategy of mutual avoidance (Dardallion, 1988; Gabor et al., 1999; Kramer-Schadt et al., 2007).

The dynamics of the matriarchal group include the isolation of pre-parturient females in spring, with their re-entry to the group with their new born, the entry of nulliparous females and the arrival of adult males during the rutting season with simultaneous departure of subadult males (Leaper et al., 1999). There is little or no aggression with the introduction of piglets to the matriarchal group (Jensen, 1986; Mauget, 1981). Subadult females split up and disperse with other subadult females to constitute a new social group, although sometimes they may return to their matriarchal group in autumn, together with their newborn piglets (Dardallion, 1988). However, it is suggested that only about 20% of yearling females may leave the matriarchal group and disperse (Kaminski et al., 2005). Adult males are frequently solitary, but subadult males

http://dx.doi.org/10.19103/AS.2020.0081.03

often roam together in small groups (Fernandez-Llario et al., 1996). Males fight for access to adult (older than one year) females and, having achieved matings, they move to another group.

Aggression is qualitatively similar in domestic pigs and their wild ancestors (Stolba, 1988). However, in contrast to domestic pigs in modern pig production systems, aggression between individuals of the wild boar in the wild and between individuals of the domestic pig returned to semi-natural settings outside of the breeding season is infrequent and rarely injurious (Stolba, 1988; Marchant-Forde and Marchant-Forde, 2005; Turner et al., 2006). Aggression is partially genetically determined and there is evidence that breeding programmes for improved growth may have resulted in increased aggression in the modern domestic pig. For example, pigs with a high breeding value for growth show more aggression when grouped with unfamiliar pigs (Rodenburg et al., 2010; Canario et al., 2012). There is also evidence that, genetically, young pigs that receive many skin injuries show improved growth performance compared to those with few injuries (Desire et al., 2015).

Social interactions are also obviously affected by the environment. Matriarchal groups of wild boar in the wild and groups of domestic pigs in semi-natural settings have a stable, linear hierarchy, based on sex and, within each sex group, the closely related factors of weight, size and strength (Mauget, 1981; Stolba and Wood-Gush, 1989). Aggression is infrequent and is generally associated with key resources such as food (Mauget, 1981; Jensen and Wood-Gush, 1984; Stolba and Wood-Gush, 1989). The inability of domestic pigs to form matriarchal groupings in modern pig production with its housing systems and management practices, such as early weaning, frequent mixing of unfamiliar pigs and penning that limits avoidance responses by pigs and dispersal and the display of appropriate submissive behaviour, obviously contributes to increased aggression. With the wide disparity between the 'natural' and 'farmed' situations, it is therefore not surprising that problems arise with aggression when domestic pigs are mixed in modern production systems (Marchant-Forde and Marchant-Forde, 2005; Marchant-Forde, 2009a).

2 Breeding and gestation of gilts and sows

2.1 Boar exposure, puberty stimulation and oestrus detection

Selected breeding gilts are generally exposed to boars for puberty stimulation at 150-180 days of age (Kraeling and Webel, 2015). Effective puberty stimulation has been recently described by several authors (Kraeling and Webel, 2015; Patterson and Foxcroft, 2019). A combination of tactile, visual, auditory and olfactory boar stimuli and daily, direct exposure to a rotation of mature boars (older than 10 months) for a minimum of 10-15 min maximizes the pubertal response in gilts. Both the stimulation of puberty and detection of puberty are facilitated by providing fence-line and direct contact (15 min daily) with multiple mature boars (Siswadi and Hughes, 1995; Patterson et al., 2016).

Direct contact with mature boars during gilt puberty stimulation and oestrus detection of both gilts and sows should be supervised to avoid stress as well as injury to gilts and sows. Based on fleeing and prolonged high-pitched vocalizations, subordinate sows in pair housing showed more fear in response to boar stimulation even when they were in oestrus, thus possibly impairing both oestrus detection and mating (Pedersen et al., 2003). Gilts and sows are also at risk of fractures or spinal cord injuries being mounted excessively by boars or other females in the pen (Levis, 2011). Rapid growth and increased body weight before mating and excessive body condition predispose gilts and young sows to lameness, and natural mating may further increase the risk of foot and leg injury (Gill, 2007). While there is no evidence, oestrus detection using full boar contact may also increase the risk of foot and leg injury in gilts and sows. Furthermore, the physical environment, particularly floor type, quality and maintenance, will affect the risk of foot and leg injury during oestrus detection and natural mating and poor natural mating conditions such as slippery floors, small floor area and obstructions such as drinkers may depress the sexual receptivity of gilts (Hemsworth et al., 1989a; Jongman et al., 1996).

2.2 Housing gilts and sows around mating and during gestation

Weaned and pregnant gilts and sows are housed under varying conditions indoors and outdoors (Marchant-Forde, 2009a; Kemp and Soede, 2012). They can be housed from weaning in gestation stalls full-time or until post-insemination or pregnancy confirmation when then housed in groups, or housed in groups from weaning. Group size can vary from small (4–5) to very large (up to 400), and the feeding systems can vary from feeding on the floor or in troughs, feeding stalls (free-access partial or free-access full-body length stalls), trickle feeders or electronic sow feeders (ESF), and thus pigs may feed simultaneously and sequentially. Space allowance can vary from 1.4 m²/pig to higher space allowances, particularly in outdoors systems. Other features in group-housing systems may also differ such as flooring, the provision and type of bedding, temperature and feeding level, although gilts and sows are mostly restrictively fed during pregnancy. These design features of the housing system can affect the welfare of gilts and sows.

Aggression is inevitable when unfamiliar pigs are mixed. Agonistic behaviour and particularly aggression may lead to injuries such as claw and limb lesions, fractures, or spinal cord damage associated with incidents such as slips, falls, knocks, bites, head slashes and toes getting stuck in the gaps of concrete slats (Anil et al., 2005). Under commercial conditions, pigs exhibit dyadic dominance relationships but mixing unfamiliar pigs of similar weight and age elicits intense aggression, particularly during the first day and is likely to be exacerbated by a close similarity in dominance ability (see review by Turner and Edwards, 2004). Pigs in small groups of 10–30 exhibit dyadic dominance relationships that show stability in a range of commercial contexts, including changes to the group composition (Turner and Edwards, 2004; Verdon and Rault, 2018). Once established, social relationships are largely maintained by low-intensity aggression, threats and avoidance behaviour (Beilharz and Cox, 1967; Gonyou, 2001). Thus, once these dyadic dominance relationships are established, there is a reduced need to deliver aggression, resulting in an overall reduction in aggression in the group (Gonyou, 2001). In larger groups, there appears to be less reliance on aggression during the period immediately post-mixing than would be expected and possible changes in social behaviour in large group sizes that may be responsible are discussed in the section on group size. Furthermore, as reviewed by Verdon and Rault (2018), once dominance relationships are established, the characteristics of the aggression changes to shorter durations but more frequent aggression and generally occurs over access for, or defence of, a resource such as food or lying space. When gilts or sows have to compete for food, aggression can remain high for at least 3–9 days after mixing (Barnett, 1997; Arey, 1999; Hemsworth et al., 2013; Verdon et al., 2016). Interestingly while frequency of aggression in gilts and sows was observed to remain high for at least 9 days after mixing, Verdon et al. (2016) found that fresh injuries were lower at 2 days than at 9 days after mixing, perhaps as a consequence of the change in the characteristics of aggression from less-frequent but longer durations to more frequent but shorter durations as the dominance relationships are established. Thus pigs in groups are exposed to a range of incidents that may cause injury such as slips, falls, knocks, bites, head slashes, collisions with other pigs and pen features and toes getting stuck in the gaps of concrete slats.

In addition to agonistic behaviour associated with the establishment of dominance relationships, agonistic behaviour can be affected by a range of housing and management practices as well as pig characteristics such as experience and genetics. As a consequence of legislative, consumer and retailer pressure on the housing of gilts and sows in gestation stalls (Matthews and Hemsworth, 2012), many countries have moved to, or are moving to, group-housing systems for gestating sows. The following section on housing around mating and gestation will therefore focus on gilts and sows housed in groups.

2.2.1 Pen design

Floor space. Floor space is a key determinant of aggression, injuries and stress when mixing and housing unfamiliar pigs. In addition to spatial requirements for physical size and basic movement, pigs need to access key resources such as feed, water and lying space. They are also motivated to interact with other pigs and to explore particularly if hungry. Therefore, they need floor space not only to access the resources but also, if necessary, to distance themselves from others, including when accessing these resources (Verdon et al., 2015a).

Floor space is particularly relevant early after mixing since insufficient space, in terms of both quantity (amount) and quality (configuration, including physical and visual barriers), can prolong aggression by affecting a pig's ability to avoid or escape others and hence delay the formation of a stable hierarchy (Lindberg, 2001). However, defining floor space allowance in terms of minimising aggression can never be definitive because, for example, the animal's response is curvilinear and the level of aggression will depend on factors such as the quality of space, access to feed and parity variation in the group.

There is evidence of increased aggression and stress in both the short and long terms if space is insufficient for gilts. Reducing floor space for gilts within the range of 1.0-3.0 m^2/animal increases aggression and stress. Floor space of 1.0 m^2/gilt elevated plasma cortisol concentrations on weeks 2 and 10 after mixing in adult gilts compared to 2.0 or 3.0 m^2/gilt (Hemsworth et al., 1986) and increased aggression around feeding on days 2-30 after mixing elevated plasma cortisol concentration and increased the cortisol response to ACTH on weeks 5 and 7 after mixing pregnant gilts compared to 1.4 and 2.0 m^2/gilt (Barnett et al., 1992; Barnett, 1997). Aggression during the first 90 min after mixing pregnant gilts was similar in floor spaces of 1 and 2 m^2/gilt (Barnett et al., 1992). However, this is in contrast to a subsequent experiment which showed reduced aggression during the first 90 min of mixing unfamiliar adult gilts when floor space was reduced from 3.4 to 1.4 m^2/gilt (Barnett et al., 1993a). The feeding system in the experiment by Hemsworth and colleagues was floor feeding and the experiments by Barnett and colleagues used a factorial design to examine both space and provision of feeding stalls.

Reducing floor space has also been shown to increase sow aggression and stress. Aggression on days 1 and 2 after mixing unfamiliar sows was higher at 1.2 than at 2.0 m^2/sow in a floor-feeding system (Taylor et al., 1997). In an experiment involving established groups of pregnant sows, each studied in each of four floor space treatments for 7-day periods and confined in individual feeding stalls around feeding, aggression on days 6 and 7 of treatment was higher at 2.0 than at 2.4, 3.6 or 4.8 m^2/sow (Weng et al., 1998). While salivary cortisol concentrations at 2 and 26 h after mixing were not affected, reducing floor space from 3.0 to 2.25 m^2/sow increased aggression on days 1, 8, 15 and 22 after mixing in a dynamic group with an electronic sow-feeding system (Remience et al., 2008). Aggression on the first two days after mixing was higher at 4.1 than at 9.3 m^2/sow in elongated rectangular and circular pens, but not in rectangular or square pens (Docking et al., 2001). Furthermore, within the range of 1.4–3.0 m²/sow, significant negative relationships were found between space allowance and both aggression at feeding and stress based on plasma cortisol concentrations at day 2 after mixing unfamiliar sows, but not at days 8–9 and 51 after mixing in a floor-feeding system (Hemsworth et al., 2013, 2016). Reducing floor space from 6.0 or 4.0 to 2.0 m²/sow did not affect aggression on days 1–4 after mixing in a floor-feeding system, but salivary cortisol concentrations were surprisingly lower at 2.0 m²/sow (Greenwood et al., 2016).

This effect of space on aggression and stress in sows found on day 2 after mixing, but not later in pregnancy (Hemsworth et al., 2013, 2016), is in contrast to that of gilts in which reducing floor space from 2.0 to 1.4 m²/gilt increased aggression on 2–30 days after mixing and elevated plasma cortisol concentration on 5 and 7 weeks after mixing (Barnett et al., 1992; Barnett, 1997). While there is no obvious explanation for these conflicting effects, it has been suggested that sows that are more experienced with group housing may adapt more quickly to spatial restriction in groups than gilts (Hemsworth et al., 2013, 2016).

Reducing floor space has been shown to increase the prevalence of skin injuries in some studies within the range of 2.0–4.0 m²/sow (Weng et al., 1998), 4.1–9.3 m²/sow (Docking et al., 2001), 1.4–3.3 m²/sow (Salak-Johnson et al., 2007) and 2.25–3.0 m²/sow (Remience et al., 2008), but not in others within the range of 1.4–3.4 m²/gilt (Barnett et al., 1993a), 1.4–3.0 m²/sow (Hemsworth et al., 2013) and 2.0–6.0 m²/sow (Greenwood et al., 2016). Differences between studies in feeding system, group size, parity and physical features in the pen may be responsible for these conflicting results on skin injuries. In an experiment examining the effects of floor space allowance (2.3, 2.8 and 3.2 m²/sow) by varying group size from 11 to 31 sows, Séguin et al. (2006a) found that skin injuries at 24 h after mixing were not affected by floor space or group size. While there was no effect of floor space on lameness during the first 3 weeks after mixing, Salak-Johnson et al. (2007) found that a floor space of 3.3 m²/sow was associated with a greater risk of lameness later in gestation than 1.4 or 2.3 m²/sow and Maes et al. (2016) suggested that sows with increased floor space may have a greater opportunity for activity which may increase the risk of injuries leading to lameness.

Floor space of sows mixed early after insemination may affect reproductive performance since stress can impair reproduction (Turner et al., 2005; Spoolder et al., 2009). Within the range of 1.4–3.0 m²/sow, a significant relationship has been shown between space allowance and farrowing rate in one experiment (Hemsworth et al., 2013), but not in another (Hemsworth et al., 2016). Although there was no time of year effect on aggression or stress in either experiment, the fertility of sows in summer in the earlier experiment was more susceptible to reduced space than in other replicates (Hemsworth et al., 2013). Unlike the earlier experiment, sows in the 2016 experiment, once confirmed pregnant at about day 27 after insemination, were housed in their treatment groups with a minimum of 1.8 m²/sow. The authors of this 2016 experiment proposed that providing more space after confirmed pregnancy for sows initially housed at 1.45 and 1.61 m²/sow in the experiment may have ameliorated spatial restriction effects on reproduction, particularly in summer. Furthermore, they suggested that a proportion of female pigs are expected to be resistant to the effects of prolonged stress or sustained increased cortisol (Turner et al., 2005).

These observations on the effects of space, particularly those on aggression and stress, indicate that a space allowance for gilts and sows of 1.4 m²/animal is likely to be too small from an animal welfare perspective and that significant improvements in gilt and sow welfare are likely to be achieved with space allowances for gilts and sows in the range of 2.0-2.4 m²/animal. Furthermore, the effects of space on aggression and stress are most pronounced soon after mixing (Hemsworth et al., 2013, 2016), highlighting the importance of floor space at mixing. With its animal welfare and economic implications, a strategy of staged-gestation penning, with more space immediately after mixing and less space later in gestation (Hemsworth et al., 2016), requires investigation. The implications of staged-gestation penning in which increased space is provided immediately after mixing and space is subsequently decreased once dominance relationships are established are considered in more detail in the section on specific-mixing pen.

Group size. There is a considerable body of research in young pigs indicating that aggression is not affected by group size ranging from 6 to 80 pigs (Turner et al., 1999, 2000; Samarakone and Gonyou, 2009). It is often assumed that housing pigs in large groups increases aggression, presumably on the basis that aggression is expected to increase in large groups because there are more dominance relationships to establish (Arey and Edwards, 1998) and that individual recognition may be difficult in large groups (Estevez et al., 2007). As reviewed by Turner and Edwards (2004), evidence suggests that in large groups there is less reliance on aggression during the immediate post-mixing phase than would be expected if the same proportion of dominance relationships is established using aggression in both large and small groups. The authors propose that this regulation of aggression in young pigs may simply be explained by the more effective avoidance of aggressive pigs that is possible in large pens with greater dimensions. However, Turner and Edwards (2004) propose that pigs may fight after mixing until a ceiling of fatigue or injury is reached or that they may become more selective over which group members they chose to fight, perhaps utilising status signals such as size.

Studies with sows indicate that while group size may not affect aggression or stress, injuries may be higher in large groups. Some limited research by Taylor et al. (1997) examining the effects of varying group sizes of 5, 10, 20 and 40 sows with a space allowance of 2.0 m²/sow has shown that the prevalence of skin injuries measured at days 5 and 53 after mixing was higher in groups of 10 than 5, although neither of these two groups sizes differed from 20 and 40. Taylor and colleagues also found a tendency for aggression to be higher in large groups at days 1 and 2 after mixing. In a large experiment, housing sows in groups of 10, 30 and 80 did not affect aggression around feeding or plasma cortisol concentrations measured at days 2, 9 and 51 after mixing (Hemsworth et al., 2013). However, group size was related to skin injuries, with groups of 10 consistently having the lowest number of total skin injuries over this period. While aggression is likely to lead to skin injuries, the authors concluded that contact with pen features associated with avoidance of other sows, particularly with fast movement in large pens, may also increase the incidence of skin injuries. Olsson et al. (1994) reported increased injuries as group size increases, but group size was confounded by space, feeding system and the presence of bedding.

Thus, other factors such as floor space and quality and competition for feed or feeding space may have a greater impact on sow aggression and stress than group size. As discussed earlier, it has been proposed that in large groups, less aggression than expected may occur because of more total space available to effectively avoid aggressive pigs in large pens with greater dimensions, pigs may fight after mixing until a ceiling of fatigue and injury is reached, and/or they become more selective over which group members they chose to fight, utilizing status signals such as size (Turner and Edwards, 2004). However, the limited number of studies on adult gilts and sows indicates that smaller groups of pigs may have less skin injuries.

Feeding system. Providing escape areas allows subordinates to avoid dominant gilts and sows (Olsson and Samuelsson, 1993) and providing feeding stalls allows gilts and sows protection from group-mates at feeding. Aggression was reduced in pens with partial stalls (0.74 m depth) during 15-90 min after mixing, but not during the first 15 min (Barnett et al., 1992), however, in a second similar study, partial stalls had no effect on aggression during the first 90 min after mixing (Barnett et al., 1993a). Differences between the two studies in floor area behind the stalls may have been responsible for this inconsistency in the results.

In comparison to floor feeding, providing group-housed pregnant gilts with feeding stalls, particularly full body-length stalls, reduces aggression, injuries (in some studies) and plasma cortisol concentrations in the long term. Feeding stalls without rear gates reduced aggression during and at 40 min after feeding on days 2-15 after mixing and reduced baseline plasma cortisol concentration and the cortisol response to ACTH on weeks 4 and 7-8 after mixing pregnant gilts but did not affect prevalence of skin injuries compared to floor feeding (Barnett et al., 1992). In a similar experiment, feeding stalls reduced aggression measured for 40 min, 10 min after feeding commenced on days 2-30 after mixing and reduced baseline plasma cortisol concentration and the cortisol response to ACTH on weeks 5 and 8 after mixing pregnant gilts compared to floor feeding (Barnett, 1997). It is of interest that a floor space of 1.0 m²/gilt, either total space or the space outside the feeding stalls, and independent of the feeding system, increases stress compared to 1.4 and 2.0 m²/gilt (Barnett et al., 1992; Barnett, 1997), highlighting the importance of floor space. Body-length stalls reduced sow aggression in established groups compared to shoulder stalls or no stalls at the feeding trough, however, this effect only occurred with dry feed and not with wet feed (Andersen et al., 1999).

Furthermore, feeding stalls, particularly full body-length stalls without rear gates, improved cell-mediated immunity in pregnant gilts in the long term (Barnett et al., 1992; Barnett, 1997). However, increased vulva biting occurred in pens with full body-length feeding stalls without rear gates (Andersen et al., 1999), indicating that stall design may influence the nature as well as the amount of aggression. As recognised by several authors, although floor feeding is competitive, gaining access to feeding stalls can also lead to competition and aggression between group-housed sows (see review by Bench et al., 2013).

Allowing group-housed-gestating sows' continuous access to feeding stalls reduces aggression and skin injuries, but not salivary cortisol in comparison to allowing access only around feeding time (Wang and Li, 2016). Thus, providing stalls allows gilts and sows protection from group-mates particularly at feeding and facilitates both access to important resources, such as feed, and avoidance responses from aggressive pigs.

While sequential feeding systems, such as an electronic sow feeding system, allow control over individual gilts and sow intake, animals have to queue to access the stalls. Although feeding orders can develop (see Spoolder et al., 2009), this can result in aggression and the displacement of subordinates at the entrance of the feeder (Bench et al., 2013). The design of the pen and the placement of the feeding stalls may affect aggression since less sows per electronic sow feeding system, a long route from exit to entry of the electronic sow feeding system and positioning the electronic sow feeding system away from busy areas or other resources (e.g. drinker, hay rack), may assist in improving accessibility to the electronic sow feeding system and reduce aggression (see review by Verdon et al., 2015a).

Flooring including bedding. Musculoskeletal problems in commercial pigs are common, occurring in all stages of commercial production and these problems can involve all components of the musculoskeletal system such as the claw, foot, lower leg, thigh, shoulder and pelvis (see reviews by Nalon et al., 2013; Pluym et al., 2013). The general clinical sign of musculoskeletal problems is lameness, an impaired movement or deviation from normal gait (Cockram and Hughes, 2011) and the main causes of lameness in pigs range from inflammation and pain (Cockram and Hughes, 2011) and infected skin and claw lesions (Velarde, 2007) to broken bones (Marchant-Forde, 2009a). Among these, it has been proposed that osteochondrosis is a major factor contributing to lameness in pigs (see review by Supakorn et al., 2018). It has been estimated that the total annual culling and death rate from 2011 to 2016 was 8.3% to 13.0% for gilts and 45.7% to 47.4% for sows, and the reported culling rate due to locomotor problems of 15.2% in the United States, 16.9% in England, 9.7% in Belgium, 15.5% in Southern Mexico, 9–15% in Sweden, Finland and Denmark, 22.5% in Southern China and 37.4% in Thailand (Supakorn et al., 2018). While claw lesions are common in sows with a prevalence ranging between 59% and 99% of sows having one or more claw lesions, not all of these sows display lameness (Heinonen et al., 2013; Pluym et al., 2011, 2013).

Lameness is a multifactorial problem affected by genetics, housing and management practices such as herd health and nutrition (Nalon et al., 2013; Pluym et al., 2013). Most lameness cases during group housing develop shortly after the sows are introduced to groups (Supakorn et al., 2018). Genetic background is implicated. Disparity in claw size, which is associated with higher culling risk, is heritable although there is a wide range of heritability values, depending on breed and statistical method (Pluym et al., 2013). Nutrition is an important predisposing factor of sow lameness, but information on the optimum feeding strategy and nutrient requirements for modern sows in contemporary housing are limited (Pluym et al., 2013; van Riet et al., 2013). Supakorn et al. (2018) have emphasised the need to fully understand the interaction between genetics and nutrition, relative to lameness and osteochondrosis.

Quality of flooring is essential to pig welfare as it is likely to have a direct effect on foot health and the culling rate from lameness (Barnett et al., 2001; Borell et al., 1997). The primary cause of lameness is pain (Cockram and Hughes, 2011) and consequently lameness is likely to be associated with some level of suffering in the animal (Dawkins, 1998). The review by Pluym et al. (2011) of research over the last 30 years on flooring and claw health in sows highlights the impact of floor quality and hygiene. For example, some of the main contributors to the development of claw lesions include group housing on fully or partly slatted floors, without bedding; wet and soiled dunging area; and old concrete flooring with rough, crumbled slat edges and enlarged gaps. Since slip-resistance, abrasiveness, hardness, surface profile and void ratio are the main floor characteristics contributing to claw health (Pluym et al., 2011, 2013; Supakorn et al., 2018), the authors recommended research on contemporary housing and flooring systems for sows to define optimum values for these floor characteristics. Foraging substrates can reduce injury caused by slipping (Heinonen et al., 2013), some abrasion is necessary for foot health (Webb and Nilsson, 1983) and the lack of natural wearing of the horn on solid floors with deep bedding may increase the risk of toe erosions, especially when straw is wet and soiled (Kilbride et al., 2009). The quality of flooring interacts with the thermal requirements and pen design on lameness in sow. For outdoor sows, site and soil type are significant factors affecting lameness (Thornton, 1990).

2.2.2 Specific mixing pen

The use of mixing pens when mixing gilts and sows has been proposed by many authors (e.g. Edwards et al., 1993; Arey and Edwards, 1998; Barnett et al., 2001; Verdon et al., 2015a; Verdon and Rault, 2018) on the basis that more floor space may provide the opportunity for avoidance by less aggressive pigs, while enabling the dominance relationships to form with less aggression, and subsequently less injuries and stress, at mixing.

The attributes of mixing pens that have been proposed to reduce aggression at mixing include: straw or rice hulls and feed, which may provide a distraction (straw or rice hulls also offers a foothold during aggression or fleeing); easy access to feed; absence of tightly confined areas in which a pig could be cornered and unable to escape from an aggressive pig; adequate space for pigs to turn around and for two pigs to easily pass side by side in all places; modifying pen size and shape since pigs require a minimum space in which they fight; the use of masking odours since anosmic pigs show reduced aggression; and sedation using pharmacological agents and grouping after dark to utilise this normal inactive time (Arey and Edwards, 1998; Barnett et al., 2001; Verdon et al., 2015a; Verdon and Rault, 2018). Although Barnett et al. (2001) concluded that all or some of these above suggestions may only be effective in postponing aggression, rather than reducing it, the aim when mixing gilts or sows should be to introduce them in a setting in which individuals can avoid aggressive ones when required with minimum risk to injury and stress while also allowing the dominance relationships to quickly form. Bedding such as straw while not likely to reduce aggression (see review by Verdon et al., 2015a) will reduce the risks of leg injury associated with aggression (see review by Spoolder et al., 2009).

As discussed in the section on floor space, one of the most obvious features of a mixing pen that will reduce aggression and stress is increased floor space (Barnett et al., 1992, 1993a). Furthermore, the sow's requirement for space may be less once the group is established (Hemsworth et al., 2013, 2016). A recent study in which the effects of housing sows at 4.0 or 6.0 m^2/sow for the first 4 days after mixing in comparison to 2.0 m^2/sow and then housing all sows at 2.0 m^2 per sow did not affect aggression, plasma cortisol concentrations or skin injuries in either period (Greenwood et al., 2016). Nevertheless, further research is required on this topic since Hemsworth et al. (2013, 2016) have shown that sows in static groups may adapt over time to a reduced space of

1.4 m²/sow and thus a strategy of staged-gestation penning, with more space immediately after insemination and less space later in gestation once the dominance relationships have been established, may address both animal welfare and economic considerations. Therefore, further research is required, particularly on the long-term effects on the pigs when subsequently placed in their gestation group systems.

A visual barrier in the pen may reduce aggression after mixing by allowing retreating pigs to visually isolate themselves from aggressive ones (Marchant-Forde and Marchant-Forde, 2005). The provision of a partial barrier centrally placed along the length of the pen reduced sow aggression in the 12-h period after mixing but not skin injuries (Edwards et al., 1993). In contrast Luescher et al. (1990) found that the provision of barriers placed perpendicular to the back of the pen and head dividers in feeding troughs did not affect the time that previously unfamiliar gilts spent fighting during 3 days after mixing. Surprisingly gilts in pens with barriers had more skin injuries on front and cheeks at day 4 after mixing, possibly arising from contact with the partitions associated with avoidance of aggressive gilts.

As discussed in the section on feeding system, aggression was reduced in pens with partial stalls early after mixing. Furthermore, full-body length feeding stalls reduced aggression around feeding on days 2–30 after mixing (Barnett et al., 1992; Barnett, 1997).

The shape of the group pen may influence aggression at mixing. Aggression in growing pigs was found to be higher in circular pens and lower in square pens, with rectangular and triangular pens exhibiting intermediate levels (Wiegand et al., 1994), leading the authors to conclude that corners may act as hide areas, which have been shown to decrease aggression in newly grouped pigs (McGlone and Curtis, 1985). There is limited evidence that aggression in adult gilts at mixing is reduced in rectangular pens compared to square pens with a floor area of 1.4 m²/gilt (Barnett et al., 1993a). However, the benefits were lost in larger pens providing 3.4 m²/gilt. In a study of pen shape (rectangular, square, elongated rectangular and circular pens) and floor space (4.1 and 9.3 m²/sow) in the first 28 hours after mixing, Docking et al. (2001) found the least aggression in the circular pen with 9.3 m²/sow.

Another management technique that has been examined is boar presence. Boar presence in the pen at mixing reduced aggression during both the feeding and non-feeding periods, and skin injuries on the first day of mixing (Docking et al., 2001). Similarly, the presence of a boar in the pen reduced the duration of fighting on days 1 and 2 after mixing during both the feeding and non-feeding periods, but not the frequency of aggression or fights (Borberg and Hoy, 2009). However, there was no effect of boar presence on skin injuries at the day after mixing. In contrast, Séguin et al. (2006b) found that the frequency of aggressive interactions and duration of fighting during the floor feeding and non-feeding periods in the first 2 days after mixing were unaffected by the presence of a boar either in the pen or with fenceline contact, although fenceline contact increased aggression during the feeding period in the 25-48 h period after mixing. Sows mixed with a boar had fewer shoulder scratches than sows without boar contact. There were also no treatment effects on agonistic behaviour for two days after boars were removed 48 h after mixing or saliva cortisol concentrations on days 1, 2 and 6 after mixing (Séguin et al., 2006b). While two of these three studies indicate that mixing with a boar provides some welfare benefits in terms of reducing sow agonistic behaviour and skin injuries, these studies are confounded by group size and floor space (Marchant-Forde and Marchant-Forde, 2005).

Some chemical interventions have been shown to reduce aggression at mixing. 'Anti-aggressive' drugs such as amperozide (Barnett et al., 1993a,b, 1996) and azaperone (Luescher et al., 1990: Csermely and Wood-Gush, 1990) appear to reduce aggression in mixed gilts in the short term for as long as the drug has efficacy, but once the effects of the drug have worn off, aggression returns to levels seen in untreated gilts.

A synthetic formulation of a maternal pheromone, which closely matches the composition of skin secretions isolated from sows, has been shown to reduce aggression and injuries in pigs mixed post-weaning (McGlone and Anderson, 2002; Guy et al., 2009). While bouts of fighting decreased in the presence of a synthetic olfactory agonist over 8 days after mixing, the overall time the sows spent involved in aggressive behaviour remained unaffected, indicating that the treated sows displayed fewer, but longer aggressive bouts. Furthermore, there were no treatment effects on injuries, salivary cortisol concentrations or conception rate. As indicated by the authors, it is not possible to determine whether the observed effects on sow behaviour were due to the 'pheromone-like' properties of the synthetic olfactory agonist, or masking by the chemical odour that was emitted by the product since anosmic pigs show reduced aggression (Luescher et al., 1990).

2.2.3 Time of mixing

Gilts and sows can be mixed at weaning, after breeding, or after they are confirmed pregnant. Higher levels of aggression and higher cortisol concentrations on the day of mixing have been observed in sows mixed in the week after insemination (early gestation) than in 5-6 weeks after insemination (Stevens et al., 2015). These effects of timing of mixing were not seen at day 7 after mixing or later, day 91 of gestation. As with cortisol and aggression, sows mixed early in gestation had more skin injuries at day 7 after mixing than sows mixed later in gestation (Stevens et al., 2015), possibly because of declining aggression. Similarly, more skin injuries, more vulva injuries and a greater incidence of lameness were observed in sows mixed early after insemination than in those mixed later in gestation (Knox et al., 2014).

In contrast to the effects on aggression and stress found by Stevens et al. (2015), Strawford et al. (2008) and Knox et al. (2014) found that aggression early after mixing was similar for sows mixed at 2–9 days after insemination and those mixed 35–46 days after insemination and Knox et al. (2014) found that sows mixed early after insemination had a smaller increase in serum cortisol from a baseline measure than did sows mixed later in gestation. Stevens et al. (2015) found no treatment effects on skin injuries at 91 days after insemination, but Li and Gonyou (2013) found that sows mixed in either static or dynamic groups early in gestation had more skin injuries before farrowing than did those mixed later in gestation. These contradictory results may be due to differences between studies in sows, such as genetics, size and experience, design features, such as feeding system, pen design and group size, and timing of measurements relative to insemination. Interestingly conception rates were lower for sows mixed early in gestation than for those mixed later in gestation or those housed in stalls for the entire gestation (Knox et al., 2014), and farrowing rates were lower for sows mixed early in gestation than for those mixed later in gestation (Li and Gonyou, 2013; Knox et al., 2014). Increased stress when mixed early after insemination may be responsible for these reproductive effects.

With the move away from stall housing of sows, there is interest in the use of grouping sows at weaning. While group housing may facilitate sexual behaviour, there is evidence that grouping may stimulate the sexual behaviour of dominant sows but suppress that of subordinate sows (Pedersen et al., 1993, 2003). Furthermore, there is evidence that oestrus may be delayed and variation in the onset of oestrus may be increased with grouping of sows at weaning (Langendijk et al., 2000; Rault et al., 2014), although with high levels of boar stimulation, neither the detection nor duration of oestrus differed between weaned sows in stalls and those in groups (Langendijk et al., 2000). Cortisol concentrations on the day after weaning were higher for sows mixed at weaning than for those housed in stalls after weaning, but cortisol concentrations were similar after insemination when the two treatment of sows were regrouped or grouped for the first time, respectively (Rault et al., 2014). The higher stress in sows mixed at weaning could be due to either the stage of reproduction or the accumulation of various stressors when weaning sows into groups (Hemsworth, 2018).

While the literature is inconsistent on the effects of the timing of mixing post-insemination, possibly because of differences between studies in intrinsic sow factors, the design of the group systems and timing of measurements, the aim should be to minimise aggression and thus injuries and stress to safeguard both sow welfare and reproduction.

2.2.4 Static and dynamic groups

Static groups are mostly formed after insemination for the entire gestation, while sows previously unfamiliar in dynamic systems are frequently introduced to the group so that they experience between 3 and 12 mixings per gestation (Marchant-Forde, 2009a). It has been suggested that aggression may be greater in dynamic groups of sows because of the frequent introduction of unfamiliar sows to the group (Arey and Edwards, 1998; Barnett et al., 2001). The few studies that have been conducted have indicated that while no effects were found on aggression, cortisol concentrations, reproductive performance and longevity, skin injuries (cuts, swellings and wounds) and lameness were greater in dynamic groups (Anil et al., 2006; Strawford et al., 2008; Li and Gonyou, 2013).

2.2.5 Hunger in gestating sows

Sows are commonly restrictively fed during gestation for productivity and lameness reasons (Meunier-Salaün et al., 2001). It has been estimated that these levels are about 60–70% of ad libitum intake (D'Eath et al., 2018) and operant conditioning studies indicate that sows fed these restricted levels can be hungry for a considerable period of the day (Hutson, 1991; Lawrence and Terlouw, 1993).

Because conventional low-fibre, high-concentrate diets require little food-searching (appetitive) behaviour and consummatory behaviours like chewing, restrictive feeding result in unfulfilled motivations to perform these natural foraging activities, leading to increased oral stereotypies (such as oral stereotypic licking, bar-biting and sham-chewing or vacuum-chewing) (Bergeron et al., 2006) (Fig. 1).

Food restriction increases stereotypies in both group- and stall-housed sows (Terlouw et al., 1991; Terlouw and Lawrence, 1993; Bergeron et al., 2000). In addition to increasing feeding levels, increasing digestible energy in these diets can also reduce stereotypies (Bergeron and Gonyou, 1997). Feeding high-fibre diets that meet the sow's nutrient requirements increases feeding time and reduces stereotypies (e.g. Robert et al., 1993; Brouns et al., 1994; Ramonet et al., 1999), but the increased feeding time appears to account for much of this reduction in stereotypy level (Robert et al., 1997).

A recent comprehensive review on mitigating hunger in pregnant sows by D'Eath et al. (2018) also concluded that providing fibre in gestating sow diets, while still restricting energy intake, can reduce redirected oral behaviours. However, fibre does not completely satisfy the motivation of sows to access food in operant tasks. Furthermore, natural foraging behaviours such as rooting and chewing suitable substrates are increased as a result of restrictive feeding, but there are foraging opportunities with the use of straw bedding or outdoor

Figure 1 A sow displaying the oral stereotypy sham-chewing.

pastures (D'Eath et al., 2018). Straw that elicits foraging behaviour has been shown to reduce oral stereotypies in gestating sows (Spoolder et al., 1995; Bergeron et al., 2006).

2.2.6 Fully or partially slatted, non-bedded and non-enriched environments for breeding pigs

Indoor production systems are considered by some to provide barren environments for animals (Barnett et al., 2001) and some authors have raised concerns about the extensive use of fully or partially slatted, non-bedded and non-enriched environments for young and breeding pigs (e.g. Marchant-Forde, 2009b). The European Union Commission Directive 2001/93/EC (European Commission Directive, 2001, p. 37) states that 'pigs must have permanent access to a sufficient quantity of material to enable proper investigation and manipulation activities, such as straw, hay, wood, sawdust, mushroom compost or a mixture of such that does not compromise the health of animals'. However, the extent to which this provides effective enrichment has been questioned (Marchant-Forde, 2009b; Van de Weerd and Day, 2009). Although the effect of space is confounded by flooring, Merlot et al. (2017, 2019) found that gestating sows housed on slatted floors with 2.4 m^2/sow of floor space had higher cortisol concentrations, granulocyte counts and immunoglobulin G than gestating sows housed on deep straw bedding with 3.5 m^2/sow. Interestingly in both experiments, sows housed on deep straw bedding with more floor space had lower preweaning mortality, implicating prenatal stress effects on the developing offspring.

The topic of enrichment and pig welfare is reviewed elsewhere in this book.

2.2.7 Individual sow characteristics

Research on sows at mixing has shown that the heritability for severe aggressive behaviours is intermediate (h^2=0.24; Løvendahl et al., 2005). The feasibility and desirability of selection for reduced aggression in pigs is considered elsewhere in this book.

Individual sow aggressive behaviour within groups of unfamiliar animals is more predictable within a gestation than between gestations (Verdon et al., 2015b; Horback and Parsons, 2016). Verdon and colleagues concluded that while genetics is likely to contribute to individual sow aggression within groups of unfamiliar animals, the reduced strength of between-gestation relationships suggests that social experience and group composition may also influence the aggressive phenotype. In fact, when housed in groups, sows may show flexibility in their use of aggression depending on the physical characteristics or behaviour of their pen mates (see reviews by Turner and Edwards, 2004; Verdon et al., 2015a). Furthermore, while variation in age exists in groups of wild boar in the wild and groups of domestic pigs in semi-natural settings and this diversity may be integral to social learning and stabilisation of the social group, little is known of the importance of social experience particularly early in life on aggression in sows (see reviews by Verdon et al., 2015a; Telkanranta and Edwards, 2017). The importance of experience is demonstrated by the findings in several species that repeated encounter and operant performance studies that intraspecific aggression can be positively reinforcing for successful aggressors, whereas the punishing effects of defeat are likely to result in avoidance (see review by Potegal, 1979).

2.3 Boar

There has been very limited research conducted on the effects of housing and management on the welfare of breeding boars. In many countries housing boars in individual stalls where they cannot turn around is now either not recommended under domestic codes of practice for pig welfare or prohibited under the regulated standards of pig welfare. For example, the European Commission Directive (2008, pg. 11) and the New Zealand Code of Welfare (Anon, 2010, pg. 21) and Canadian Code of Practice (Anon, 2014, pg. 13) require for all holdings that boar pens must provide sufficient space to allow the boar to turn around. The Australian Model Code of Practice requires that boars accommodated individually in stalls must be able to stand, get up and lie down without being obstructed by the bars and fittings of the stall, lie with limbs extended to stretch and to be able to freely undertake such movements

(Anon, 2010, pg. 8). Furthermore, this code requires that if boars are housed continuously in stalls, they must be released for mating or exercise at least twice per week. Recommendations for floor space for boars in individual pens varies from 5.6 m² for partially or fully slatted floors (Anon, 2014, pg. 51), and 6.0 m² (European Commission Directive, 2008, pg. 11; Anon, 2010, pg. 8) and 7.4 m² for solid bedded floors (Anon, 2014, pg. 51). The New Zealand Code of Welfare (Anon, 2010) provides no minimum floor space, but requires that boars must be provided with sufficient space so that they can stand up, turn around and lie comfortably in a natural position.

Artificial insemination (AI) is widely used in modern commercial pig production and boars are mainly housed in stalls in AI centres, but pens are also used (Singleton, 2001; Knox, 2016). Surprisingly, some codes of practice or standards such as those in Australia (Anon, 2010) and the EU (European Commission Directive, 2008) do not specifically refer to housing of AI boars, and others such as in Canada (Anon, 2014) have a disclaimer that the code does not apply to associated industries such as AI boar stud farms.

Adult males of the wild ancestors of the domestic pig are frequently solitary in the wild, but during the breeding season fight to mate adult females in matriarchal groups and then move to another group (Fernandez-Llario et al., 1996). Thus, individual housing per se with sexual activity associated with regular contact with females when used for oestrus detection and mating in commercial production units or regular semen collection in AI centres, is unlikely to be stressful, although the effects of using boars solely for oestrus detection without mating on their sexual frustration and stress have not been examined.

The post-pubertal social environment of the boar appears to affect the sexual behaviour of boars. Isolation of mature boars from female pigs depresses their sexual behaviour (see review by Hemsworth and Tilbrook, 2007); however, this effect is not permanent since rehousing isolated boars near females restores their sexual behaviour within 4 weeks. Interestingly, the oestrous status of the females did not influence the effectiveness of females in maintaining the sexual behaviour of mature boars. The effect on sexual behaviour of using boars solely for oestrus detection in commercial production units or semen collection in AI centres is unknown, although low sexual motivation of AI boars may be less critical because of the moderate collection frequency generally required (Singleton, 2001; Knox, 2016). However, data from AI centres indicate that 15% of boars in three 250-head boar studs in the United States could not be trained to collection on a dummy (Flowers, 2008) and apart from boars with either poor semen or reduced for their semen, 21% of boars at a Polish AI centre from 1998 to 2013 were culled for poor sexual motivation.

While group housing of breeding boars is common in outdoor systems, boars are generally housed individually in indoor production systems. Boars housed in groups are reported to have better physical conditions, presumably because of improved space for exercise (Cordoba-Dominguez et al., 1991). In a study examining the effects of housing mature boars individually either in stalls (0.8 × 2.9 m) housed adjacent to a mating pen compared or in pens (2.6 × 3m) away from the mating pen, Levis et al. (1995) found no treatment effects on daytime baseline concentrations of plasma cortisol or the cortisol response to administration of ACTH. These physiological measures of stress provide no evidence that housing boars individually in stalls adjacent to a mating pen has any adverse effects on their welfare relative to housing boars individually in pens. However, in light of the effects of housing gilts and sows in stalls during gestation, it would be prudent to examine the welfare implications of stall housing of boars used for natural matings, boars solely used for puberty stimulation and oestrus detection and boars housed in AI centres. There is evidence that housing gestating gilts and sows in stalls depresses their welfare. Pregnant gilts in stalls experienced a chronic stress response, based on a sustained elevation of basal cortisol concentrations, similar to that seen in tethered gilts, and active avoidance of neighbours (see review by Barnett et al., 2001) as well as a higher incidence of lameness and an increased neutrophil to lymphocyte ratios, perhaps as a consequence of increased stress (Karlen et al., 2007).

Lameness and penile injuries may not physically allow the achievement of copulation or may inhibit copulation because of pain (Christensen, 1953), while injury sustained during copulation may produce a psychological effect for some time after physical recovery has occurred, again inhibiting copulation. Prevention of lameness and penile injuries should include attention to the design and maintenance of the accommodation and mating or semen collection areas, appropriate supervision and assistance at mating or semen collection, and selection for heritable conformational traits of the feet and legs (Hemsworth and Tilbrook, 2007). Obesity and skeletal defects can also affect the sexual behaviour of boars, often by affecting mobility and thus mating competency. Lameness or locomotor problems, and death and euthanasia account for 8-12% and 7-8%, respectively, of removals of breeding boars in commercial production units (D'Allaire and Leman, 1990; Koketsu and Sasaki, 2009) and lameness is one of the main reasons for culling boars from AI centres (Knecht et al., 2017). Lameness or locomotor injuries and death in AI and commercial production units are likely to be painful and thus a welfare concern (Whay et al., 2003).

While it does not appear to be widely viewed as a high priority based on stall housing of boars in some countries, research is required on the welfare implications of stall housing of boars used for natural matings, boars solely used for puberty stimulation and oestrus detection and boars housed in AI centres.

3 Animal management

3.1 Animal abuse and cruelty

It is indisputable that there have been cases of animal abuse and cruelty in livestock production, just as there have been cases involving companion animals, and these are likely to continue. Animal welfare is a highly emotive subject and the news media frequently report animal welfare abuses, often as a result of covert recording or filming by animal rights organisations. These stories generate considerable public interest and the message conveyed is that the livestock industry in question engages in poor animal welfare practices and that some aspect of it should be substantially changed or even banned.

Anecdotal experience in extensively studying human-animal interactions in the livestock industries is that the prevalence of deliberate mistreatment of animals appears to be low (Hemsworth and Coleman, 2011). That is not to say that there are occasions when an individual might mistreat an animal because of a particular set of circumstances, such as in response to the animal's behaviour or because the person is upset for another reason or is unwell (Coleman and Hemsworth, 2014). There also will be cases where a person is so insensitive to the state of the animal that their behaviour is persistently inappropriate. However, most of the instances of poor behaviour appear to arise because the management of animals has largely been learned 'on the job' and people may be unaware of the effect of their behaviour on the welfare of the animals under their care.

3.2 Stockmanship

Management affects animal welfare in several ways (Hemsworth and Coleman, 2009, 2011). At the level of farm management, human resource management practices, including employee selection and training, and animal management practices, such as best practice in housing and husbandry, and implementation of welfare protocols and audits, all impact on farm animal welfare. At the stockperson level, together with the opportunity to perform their tasks well, stockpeople require a range of well-developed husbandry skills and knowledge to effectively care for and manage farm animals.

Stockpeople clearly require a basic knowledge of both the requirements and behaviour of farm animals, and also must possess a range of well-developed husbandry and management skills to care for and manage their animals effectively. Other important job-related characteristics include job satisfaction, work motivation and motivation to learn new skills and knowledge about the animal. Since the 1980s there has been an ever-increasing body of evidence on the profound effects of stockperson interactions on farm animal fear and in turn, animal stress, welfare and productivity. Research in the livestock industries, including the pig industry and supported by handling studies in laboratories, demonstrates sequential relationships between stockperson attitudes to interacting and working with their animals, stockperson interactions and animal fear, stress and productivity (see Coleman, 2004; Hemsworth and Coleman, 2011). The specific attitudes identified in the field studies were stockperson attitudes towards their animals and working with them, their beliefs about other people's expectations of them, and their beliefs about the extent to which they have control over how they interact with and manage their animals. These, in turn, determine the nature and extent of stockperson interactions with their animals and, in situations in which stockperson behaviour is poor, animal fear and stress are elevated and thus both animal productivity and welfare are at risk. For example, it has been shown that positive attitudes among stockpeople to the use of patting and talking and to the avoidance of intense verbal and physical effort to handle gilts and sows are negatively correlated with the use of negative tactile interactions such as slaps, pushes and hits in handling pigs (Hemsworth et al., 1989b; Coleman et al., 1998). In turn, the frequent use of negative tactile interactions was positively associated with increased fear of humans. Furthermore, attitudes to handling pigs are also correlated with aspects of the job such as job satisfaction, work motivation and motivation to learn new skills and knowledge about the pig (Coleman et al., 1998). Therefore, attitudes influence not only the manner in which stockpeople handle pigs, but also their motivation to care for pigs.

3.3 Minimising handling stress

This research has shown that the way in which stockpeople routinely handle their animals is largely learned and that a major factor affecting stockperson behaviour is stockperson attitudes. These attitudes are based on the beliefs that stockpeople hold about the animals and are shaped by both ad hoc advice from co-workers and by direct and indirect experiences with animals. They are therefore learnt, and consequently training of stockpeople targeting the key attitudes and related behaviour that affect animal fear and stress should be effective in reducing animal fear and stress and consequently minimising risks to animal welfare through poor handling. In fact, when training programmes that are based on an understanding of both the beliefs that underpin behaviour and the appropriate ways of handling animals are used, there is a dramatic and sustained improvement in stockperson behaviour with a concomitant improvement in animal welfare as well as animal productivity (Hemsworth and Coleman, 2011; Coleman and Hemsworth, 2014). Indeed, studies in the pig industry (Hemsworth et al., 1994a; Coleman et al., 2000) have shown that cognitive behavioural training, in which these key attitudes and behaviour of stockpeople are targeted, can improve the attitudes and behaviour of stockpeople towards their pigs successfully, with consequent beneficial effects on gilt and sow fear and reproduction (Fig. 2).

Thus, training programmes that target the attitude and behaviour of stockpeople offer considerable opportunity to improve pig welfare. Improved human-animal interactions may also enhance job-related characteristics, such as job satisfaction, motivation and commitment, thereby potentially improving the stockperson's job performance and career prospects (Coleman and Hemsworth, 2014). Thus, including training targeting the attitudes and behaviours of stockpeople towards pigs in conjunction with the technical skills, and knowledge of stockpeople is likely to not only reduce the stress associated with handling and husbandry procedures involving humans, but also improve the motivation in stockpeople to learn new technical skills and knowledge and to apply these competencies in the management of the animals under their care.

Most industries have animal welfare policies in place, but policies alone do not change the behaviour of individual stockpeople unless there are major sanctions that are enforced but this requires constant surveillance which is rarely acceptable in the workplace. Direct interventions, in the form of targeted behaviour change training programmes, as research has shown, provide a proactive means to retain a skilled workforce that acts in the best interests of the animals under their care (Coleman and Hemsworth, 2019).

4 Conclusion and future trends

Aggression and consequently stress are inevitable when unfamiliar pigs are mixed. Agonistic behaviour and particularly aggression may also lead to injuries

Figure 2 Positive attitudes to handling pigs are associated with positive handling of pigs.

such as claw and limb lesions, fractures or spinal cord damage associated with incidents such as slips, falls, knocks, bites, head slashes and toes getting stuck in the gaps of concrete slats. Pen design is an important determinant of the welfare of group-housed gilts and sows. Floor space is particularly relevant early after mixing since insufficient space, in terms of both quantity (amount) and quality (configuration, including physical and visual barriers), can prolong aggression by affecting a pig's ability to avoid or escape other pigs and hence delay the formation of a stable hierarchy. Providing group-housed pregnant pigs with feeding stalls, particularly full body-length stalls, reduces the long-term aggression, plasma cortisol concentrations and often injuries. Quality of flooring in terms of bedding, dry dunging area, and well-maintained and designed slatted and solid floors is essential to pig welfare as it is likely to have a direct effect on foot health and the culling rate from lameness. While aggression, cortisol concentrations, reproductive performance and longevity are not affected, skin injuries and lameness are greater in dynamic than static groups.

While not studied extensively, particularly in sows, the use of mixing pens in which more space, free-access, full-body length stalls, visual barriers and bedding are provided is likely to provide the opportunity for avoidance by less-aggressive gilts and sows, enabling the dominance relationships to form with less aggression, and subsequently less injuries and stress. Further research however is required, particularly on the long-term effects on the pigs when subsequently placed in gestation group systems.

Although there is evidence in other species that early experience may affect social skills later in life, there are few studies on the effects of early social experience on aggressive behaviour of sows. Genetic selection also has the potential to reduce aggression, and therefore, continued research on the opportunity to genetically select against aggressiveness and its broader implications is required.

While it does not appear to be widely viewed as a high priority, research is required on the welfare implications of stall housing of boars used for natural matings, boars solely used for puberty stimulation and oestrus detection and boars housed in AI centres.

Lameness is a multifactorial problem affected by genetics, housing and management practices such as herd health and nutrition, and prevention of lameness from developing in group-housed sows is a major challenge since most lameness during group housing developing shortly after the sows are introduced to groups. The quality of flooring is essential to pig welfare as it is likely to have a direct effect on foot health and the culling rate from lameness. Slip-resistance, abrasiveness, hardness, surface profile and void ratio are the main floor characteristics contributing to lameness and research on contemporary housing and flooring systems is required to define optimum values for these floor characteristics. The interaction between genetics and nutrition, relative to lameness and osteochondrosis needs to be fully understood.

Direct contact with mature boars during gilt puberty stimulation and oestrus detection of both gilts and sows should be supervised to avoid stress as well as injury to gilts and sows. The physical environment, particularly floor type, quality and maintenance, will contribute to the risk of foot and leg injury during puberty stimulation, oestrus detection and natural mating.

Most industries have animal welfare policies in place, but policies alone do not change the behaviour of individual stockpeople. Direct interventions, in the form of targeted behaviour change training programmes, as research has shown, provide a proactive means to retain a skilled workforce that acts in the best interests of the animals under their care. It is likely that general community will place an increasing emphasis on ensuring the competency of stockpeople to manage the welfare of livestock.

In conclusion, animal welfare problems for pigs during breeding and gestation are generally less a function of the type of housing system than of how well it operates. The focus on different housing systems ignores many of the important factors that can affect animal welfare such as the design of the system per se and the quality of management, particularly stockmanship. As persuasively stated by Fraser (2005, pp. 8C, 9C):

> some of the most important determinants of animal welfare are not specific to any one type of housing and production system. Whether dairy cows have more health problems in tie-stalls or on pasture is a matter of debate, but there is no disagreement that their welfare is improved by staff who are skilled at detecting and treating disease. Whether sows are better off in stalls or group-housed systems is sometimes disputed, but there is no disagreement that good maintenance and functioning of equipment is important to their welfare. In fact, if we think of animal welfare as being influenced by such key factors such as staff time and skill, substrate, temperature, feed quality and disease prevention measures, then animal welfare problems may be less a function of the type of rearing system - confinement, semi-confinement or extensive - than of how well it operated.

5 Where to look for further information

The welfare of breeding pigs has been reviewed by Barnett et al. (2001), Marchant-Forde (2009a) and Verdon et al. (2015a). The topic of human-animal interactions in the livestock industries has been reviewed by Hemsworth and Coleman (2011) and Coleman and Hemsworth (2014), and details of the commercialized training programme targeting attitudinal and behavioural change in stockpeople ('ProHand') are available at: https://www.animalwe lfare-science.net/uploads/1/2/3/2/123202832/prohand_background_jul19 .pdf.

6 References

Andersen, I. L., Bøe, K. E. and Kristiansen, A. L. (1999). The influence of different feeding arrangements and food type on competition at feeding in pregnant sows. *Appl. Anim. Behav. Sci.* 65(2): 91–104.

Anil, L., Anil, S. S., Deen, J., Baidoo, S. K. and Walker, R. D. (2006). Effect of group size and structure on the welfare and performance of pregnant sows in pens with electronic sow feeders. *Can. J. Vet. Res.* 70(2): 128–136.

Anil, S. S., Anil, L., Deen, J., Baidoo, S. K. and Wheaton, J. E. (2005). Characterization of claw lesions in gestating sows. In: 2005 Allen D. Leman Swine Conference, UDA, Minneapolis, MN, pp. 193–199.

Anon. (2010). *Animal Welfare (Pigs), Code of Welfare: A Code of Welfare Issued under the Animal Welfare Act 1999*. National Animal Welfare Advisory Committee, New Zealand.

Anon. (2014). *Code of Practice for the Care and Handling of Pigs, Canada*. National Farm Animal Care Council (NFACC), Canada.

Arey, D. S. (1999). Time course for the formation and the disruption of social organisation in group-housed sows. *Appl. Anim. Behav. Sci.* 62(2–3): 199–207.

Arey, D. S. and Edwards, S. A. (1998). Factors influencing aggression between sows after mixing and the consequences for welfare and production. *Livest. Prod. Sci.* 56(1): 61–70.

Barnett, J. L. (1997). Modifying the design of group pens with individual feeding places affects the welfare of pigs. In: Bottcher, R. W. and Hoff, S. J. (Eds). Proceedings of the 5th International Symposium American Society of Agricultural Engineers, St. Joseph, MI, pp. 613–618.

Barnett, J. L., Hemsworth, P. H., Cronin, G. M., Jongman, E. C. and Hutson, G. D. (2001). A review of the welfare issues for sows and piglets in relation to housing. *Aust. J. Agric. Res.* 52(1): 1–28.

Barnett, J. L., Hemsworth, P. H., Cronin, G. M., Newman, E. A., McCallum, T. H. and Chilton, D. (1992). Effects of pen size, partial stalls and method of feeding on welfare-related behavioural and physiological responses of group-housed pigs. *Appl. Anim. Behav. Sci.* 34(3): 207–220.

Barnett, J. L., Cronin, G. M., McCallum, T. H. and Newman, E. A. (1993a). Effects of pen size/shape and design on aggression when grouping unfamiliar adult pigs. *Appl. Anim. Behav. Sci.* 36(2–3): 111–122.

Barnett, J. L., Cronin, G. M., McCallum, T. H. and Newman, E. A. (1993b). Effects of 'chemical intervention' techniques on aggression and injuries when grouping unfamiliar adult pigs. *Appl. Anim. Behav. Sci.* 36(2–3): 135–148.

Barnett, J. L., Cronin, G. M., McCallum, T. H., Newman, E. A. and Hennessy, D. P. (1996). Effects of grouping unfamiliar adult pigs after dark, after treatment with amperozide and by using pens with stalls, on aggression, skin lesions and plasma cortisol concentrations. *Appl. Anim. Behav. Sci.* 50(2): 121–133.

Beilharz, R. G. and Cox, D. F. (1967). Social dominance in swine. *Anim. Behav.* 15(1): 117–122.

Bench, C. J., Rioja-Lang, F. C., Hayne, S. M. and Gonyou, H. W. (2013). Group gestation sow housing with individual feeding–II: how space allowance, group size and composition, and flooring affect sow welfare. *Livest. Sci.* 152(2–3): 218–227.

Bergeron, R., Badnell-Waters, A. J., Lambton, S. and Mason, G. (2006). Stereotypic oral behaviour in captive ungulates: foraging, diet and gastrointestinal function. In:

Mason, G. and Rushen, J. (Eds). *Stereotypic Animal Behaviour: Fundamentals and Applications to Welfare* (2nd edn.). CABI Publishing, Wallingford, UK, pp. 19-27.

Bergeron, R., Bolduc, J., Ramonet, Y., Meunier-Salaün, M. C. and Robert, S. (2000). Feeding motivation and stereotypies in pregnant sows fed increasing levels of fibre and/or food. *Appl. Anim. Behav. Sci.* 70(1): 27-40.

Bergeron, R. and Gonyou, H. W. (1997). Effects of increasing energy intake and foraging behaviours on the development of stereotypies in pregnant sows. *Appl. Anim. Behav. Sci.* 53(4): 259-270.

Borberg, C. and Hoy, S. (2009). Mixing of sows with or without the presence of a boar. *Livest. Sci.* 125(2-3): 314-317.

Borell, E., von, B. D. M., Csermely, D., Dijkhuizen, A. A., Edwards, S. A., Jensen, P., Madec, F. and Stamataris, C. (1997). The welfare of intensively kept pigs. In: Report of the Scientific Veterinary Committee, European Union, Brussels, Belgium.

Brouns, F., Edwards, S. A. and English, P. R. (1994). Effect of dietary fibre and feeding system on activity and oral behaviour of group housed gilts. *Appl. Anim. Behav. Sci.* 39(3-4): 215-223.

Canario, L., Turner, S. P., Roehe, R., Lundeheim, N., D'Eath, R. B., Lawrence, A. B., Knol, E., Bergsma, R. and Rydhmer, L. (2012). Genetic associations between behavioural traits and direct-social effects of growth rate in pigs. *J. Anim. Sci.* 90(13): 4706-4715.

Christensen, N. O. (1953). Impotentia coeundi in boars due to arthrosis deformans. In: 15th International Veterinary Congress, Part I, Stockholm, Sweden, Vol. 2, pp. 742-745; Part II, pp. 332-333.

Cockram, M. S. and Hughes, B. (2011). Health and disease. In: Appleby, M. C., Hughes, B. and Mench, J. (Eds). *Animal Welfare*. CABI Publishing, Wallingford, UK, pp. 120-137.

Coleman, G. J. (2004). Personnel management in livestock industries. In: Benson, G. J. and Rollin, B. E. (Eds). *The Well-Being of Farm Animals: Challenges and Solutions.* Blackwell Publishing, Ames, IA, pp.167-181.

Coleman, G. J. and Hemsworth, P. H. (2014). Training to improve stockperson beliefs and behaviour towards livestock enhances welfare and productivity. *Scientific and Technical Review of the Office International des Epizooties* 33: 131-137.

Coleman, G. J. and Hemsworth, P. H. (2019). Scientific commentary: animal abuse and cruelty in livestock production. Animal Welfare Science Centre. Available at: https ://www.animalwelfare-science.net/news/scientific-commentary-animal-abuse-and -cruelty-in-livestock-production; https://www.animalwelfare-science.net/ (Accessed 20th September 2019).

Coleman, G. J., Hemsworth, P. H., Hay, M. and Cox, M. (1998). Predicting stockperson behaviour towards pigs from attitudinal and job-related variables and empathy. *Appl. Anim. Behav. Sci.* 58(1-2): 63-75.

Coleman, G. J., Hemsworth, P. H., Hay, M. and Cox, M. (2000). Modifying stockperson attitudes and behaviour towards pigs at a large commercial farm. *Appl. Anim. Behav. Sci.* 66(1-2): 11-20.

Cordoba-Dominguez, J., Dunne, J., Cliff, A., MacPherson, O., Menaya, C., Vidal, J. and English, P. (1991). Evaluation of the behaviour of mature boars housed in groups of four. In: Proceedings of the Winter Meeting British Society of Animal Production, Scarborough, UK, pp. 65.

Csermely, D. and Wood-Gush, D. G. M. (1990). Agonistic behaviour in grouped sows. II. How social rank affects feeding and drinking behaviour. *Ital. J. Zool.* 57(1): 55-58.

D'Allaire, S. and Leman, A. D. (1990). Boar culling in swine breeding herds in Minnesota. *Can. Vet. J.* 31(8): 581-583.

Dardallion, M. (1988). Wild boar social groupings and their seasonal changes in the Camargue, southern France. *Z. Säugetierkd.* 53: 22-30.

Dawkins, M. S. (1998). Evolution and animal welfare. *Q. Rev. Biol.* 73(3): 305-328.

D'Eath, R. B., Jarvis, S., Baxter, E. M. and Houdijk, J. (2018). Mitigating hunger in pregnant sows. In: Spinka, M. (Ed.). *Advances in Pig Welfare, Woodhead Publishing Series in Food Science, Technology and Nutrition.* Woodhead Publishing, Duxford, UK.

Desire, S., Turner, S. P., D'Eath, R. B., Doeschl-Wilson, A. B., Lewis, C. R. G. and Roehe, R. (2015). Genetic associations of short- and long-term aggressiveness identified by skin lesions with growth, feed efficiency and carcass characteristics in growing pigs. *J. Anim. Sci.* 93(7): 3303-3312.

Docking, C. M., Kay, R. M., Day, J. E. L. and Chamberlain, H. L. (2001). The effect of stocking density, group size and boar presence on the behaviour, aggression and skin damage of sows mixed in a specialized mixing pen at weaning. In: Proceedings of the British Society of Animal Science, Scarborough, UK, p. 46 (Abst).

Edwards, S. A., Mauchline, S. and Stewart, A. H. (1993). Designing pens to minimise aggression when sows are mixed. *Farm Build. Prog.* 113: 20-23.

Estevez, I., Andersen, I. and Naevdal, E. (2007). Group size, density and social dynamics in farm animals. *Appl. Anim. Behav. Sci.* 103(3-4): 185-204.

European Commission Directive (2008). Council Directive 2008/120/EC of 18 December 2008 laying down minimum standards for the protection of pigs. *Official Journal* 18.2.2009 L 47/5-13.

Fernandez-Llario, P., Carranza, J. and Hidalgo de Trucios, S. J. (1996). Social organization of the wild boar (*Sus scrofa*) in Donana National Park. *Misc. Zool.* 19: 9-18.

Fraser, D. (2005). Animal welfare and the intensification of animal production. An alternative interpretation. *FAO Readings in Ethics* 2. Editorial Production and Design Group, Publishing Management Service, FAO, Rome. ISBN 92-5-105386-3.

Gabor, T. M., Hellgren, E. C., van den Bussche, R. A. and Silvy, N. J. (1999). Demography, sociospatial behaviour and genetics of feral pigs (*Sus scrofa*) in a semi-arid environment. *J. Zool.* 247(3): 311-322.

Gill, P. (2007). Nutritional management of the gilt for lifetime productivity – feeding for fitness or fatness? In: *Today's Challenges. Tomorrow's Opportunities*, London Swine Conference Proceedings, London, Ontario, Canada, pp. 83-99.

Gonyou, H. W. (2001). The social behaviour of pigs. In: Keeling, L. J. and Gonyou, H. W. (Eds). *Social Behaviour in Farm Animals*. CAB International, Wallingford, UK, pp. 147-168.

Greenwood, E. C., Plush, K. J., van Wettere, W. H. E. J. and Hughes, P. E. (2016). Group and individual sow behaviour is altered in early gesation by space allowance in the days immediately following grouping. *J. Anim. Sci.* 94(1): 385-393.

Guy, J. H., Burns, S. E., Barker, J. M. and Edwards, S. A. (2009). Reducing post-mixing aggression and skin lesions in weaned pigs by application of a synthetic maternal pheromone. *Anim. Welf.* 18: 249-255.

Heinonen, M., Peltoniemi, O. and Valros, A. (2013). Impact of lameness and claw lesions in sows on welfare, health and production. *Livest. Sci.* 156(1-3): 2-9.

Hemsworth, P. H. (2018). Key determinants of pig welfare: implications of animal management and housing design on livestock welfare. *Anim. Prod. Sci.* 58(8): 1375-1386.

Hemsworth, P. H., Barnett, J. L., Coleman, G. J. and Hansen, C. (1989b). A study of the relationships between the attitudinal and behavioural profiles of stockpersons and the level of fear of humans and reproductive performance of commercial pigs. *Appl. Anim. Behav. Sci.* 23(4): 301–314.

Hemsworth, P. H., Barnett, J. L., Hansen, C. and Winfield, C. G. (1986). Effects of social environment on welfare status and sexual behaviour of female pigs. II. Effects of space allowance. *Appl. Anim. Behav. Sci.* 16(3): 259–267.

Hemsworth, P. H. and Coleman, G. J. (2009). Animal welfare and management. In: Smulders, F. J. M. and Algers, B. (Eds). *Food Safety Assurance and Veterinary Public Health Volume 5. Welfare Productions Animals: Assessment and Management Risks.* Wageningen Academic Publishers, the Netherlands, pp. 133–147.

Hemsworth, P. H. and Coleman, G. J. (2011). *Human-Livestock Interactions: The Stockperson and the Productivity and Welfare of Farmed Animals* (2nd edn.). CAB International, Oxon, UK.

Hemsworth, P. H., Coleman, G. J. and Barnett, J. L. (1994). Improving the attitude and behaviour of stockpersons towards pigs and the consequences on the behaviour and reproductive performance of commercial pigs. *Appl. Anim. Behav. Sci.* 39(3–4): 349–362.

Hemsworth, P. H., Hansen, C. and Winfield, C. G. (1989a). The influence of mating conditions on the sexual behaviour of male and female pigs. *Appl. Anim. Behav. Sci.* 23(3): 207–214.

Hemsworth, P. H., Morrison, R. S., Tilbrook, A. J., Butler, K. L., Rice, M. and Moeller, S. J. (2016). Effects of varying floor space on aggressive and cortisol concentrations in group housed sows. *J. Anim. Sci.* 94(11): 4809–4818.

Hemsworth, P. H., Rice, M., Nash, J., Giri, K., Butler, K. L., Tilbrook, A. J. and Morrison, R. S. (2013). Effects of group size and floor space allowance on grouped sows: aggression, stress, skin injuries and reproductive performance. *J. Anim. Sci.* 91(10): 4953–4964.

Hemsworth, P. H. and Tilbrook, A. J. (2007). Sexual behavior of male pigs. *Horm. Behav.* 52(1): 39–44.

Horback, K. M. and Parsons, T. D. (2016). Temporal stability of personality traits in group-housed gestating sows. *Animal* 10(8): 1351–1359.

Hutson, G. D. (1991). A note on hunger in the pig: sows on restricted rations will sustain an energy deficit to gain additional food. *Anim. Sci.* 52(1): 233–235.

Jensen, P. (1986). Observations on the maternal behaviour of free-ranging domestic pigs. *Appl. Anim. Behav. Sci.* 16(2): 131–142.

Jensen, P. and Wood-Gush, D. G. M. (1984). Social interactions in a group of free-ranging sows. *Appl. Anim. Behav. Sci.* 12(4): 327–337.

Jongman, E. C., Hemsworth, P. H. and Galloway, D. B. (1996). The influence of conditions at the time of mating on the sexual behaviour of male and female pigs. *Appl. Anim. Behav. Sci.* 48(3–4): 143–150.

Kaminski, G., Brandt, S., Baubet, E. and Baudoin, C. (2005). Life-history patterns in female wild boars (Sus scrofa): mother–daughter postweaning associations. *Can. J. Zool.* 83(3): 474–480.

Karlen, G. A. M., Hemsworth, P. H., Gonyou, H. W., Fabrega, E., Strom, A. D. and Smits, R. J. (2007). The welfare of gestating sows in conventional stalls and large groups on deep litter. *Applied Animal Behaviour Science* 105(1–3): 87–101.

Kemp, B. and Soede, N. M. (2012). Reproductive issues in welfare-friendly housing systems in pig husbandry: a review. *Reprod. Domest. Anim.* 47(Suppl. 5): 51–57.

Kilbride, A., Gillman, C., Ossent, P. and Green, L. (2009). Impact of flooring on the health and welfare of pigs. *In Practice* 31(8): 390-395.

Knecht, D., Jankowska-Mąkosa, A. and Duziński, K. (2017). Analysis of the lifetime and culling reasons for AI boars. *J. Anim. Sci. Biotechno.* 8: 49.

Knox, R., Salak-Johnson, J., Hopgood, M., Greiner, L. and Connor, J. (2014). Effect of day of mixing gestating sows on measures of reproductive performance and animal welfare. *J. Anim. Sci.* 92(4): 1698-1707.

Knox, R. V. (2016). Artificial insemination in pigs today. *Theriogenology* 85(1): 83-93.

Koketsu, Y. and Sasaki, Y. (2009). Boar culling and mortality in commercial swine breeding herds. *Theriogenology* 71(7): 1186-1191.

Kraeling, R. R. and Webel, S. K. (2015). Current strategies for reproductive management of gilts and sows in North America. *J. Anim. Sci. Biotechnol.* 6(1): 3.

Kramer-Schadt, S., Fernández, N. and Thulke, H.-H. (2007). Potential ecological and epidemiological factors affecting the persistence of classical swine fever in wild boar *Sus scrofa* populations. *Mamm. Rev.* 37(1): 1-20.

Langendijk, P., Soede, N. M. and Kemp, B. (2000). Effects of boar contact and housing conditions on estrus expression in weaned sows. *J. Anim. Sci.* 78(4): 871-878.

Lawrence, A. B. and Terlouw, E. M. C. (1993). A review of behavioral factors involved in the development and continued performance of stereotypic behaviors in pigs. *J. Anim. Sci.* 71(10): 2815-2825.

Leaper, R., Massei, G., Gorman, M. L. and Aspinall, R. (1999). The feasibility of reintroducing wild boar (Sus scrofa) to Scotland. *Mamm. Rev.* 29(4): 239-258.

Levis, D. G., Barnett, J. L., Hemsworth, P. H. and Jongman, E. C. (1995). The effect of breeding facility and sexual stimulation on plasma cortisol in boars. *J. Anim. Sci.* 73(12): 3705-3711.

Li, Y. Z. and Gonyou, H. W. (2013). Comparison of management options for sows kept in pens with electronic feeding stations. *Can. J. Anim. Sci.* 93(4): 445-452.

Lindberg, A. C. (2001). Group life. In: Keeling, L. J. and Gonyou, H. W. (Eds). *Social Behaviour in Farm Animals*. CABI Publishing, Oxfordshire, UK, pp. 37-58.

Løvendahl, P., Damgaard, L. H., Nielsen, B. L., Thodberg, K., Su, G. and Rydhmer, L. (2005). Aggressive behaviour of sows at mixing and maternal behaviour are heritable and genetically correlated traits. *Livest. Prod. Sci.* 93(1): 73-85.

Luescher, U. A., Friendship, R. M. and McKeown, D. B. (1990). Evaluation of methods to reduce fighting among regrouped gilts. *Can. J. Anim. Sci.* 70(2): 363-370.

Maes, D., Pluym, L. and Peltoniemi, O. (2016). Impact of group housing of pregnant sows on health. *Porcine Health Manag.* 2: 17.

Marchant-Forde, J. N. (2009a). Welfare of dry sows. In: Marchant-Forde, J. N. (Ed). *The Welfare of Pigs*. Springer Science and Business Media, New York City, pp. 95-139.

Marchant-Forde, J. N. (2009b). Future perspectives of the welfare of pigs. In: Marchant-Forde, J. N. (Ed). *The Welfare of Pigs*. Springer Science and Business Media, New York City, pp. 331-342.

Marchant-Forde, J. N. and Marchant-Forde, R. M. (2005). Minimizing inter-pig aggression during mixing. *Pig News Inf.* 26: 63N-71N.

Matthews, L. R. and Hemsworth, P. H. (2012). Drivers of change: law, international markets, and policy. *Anim. Front.* 2(3): 40-45.

Mauget, R. (1981). Behavioural and reproductive strategies in wild forms of *Sus scrofa* (European wild boar and feral pigs). In: Sybesma, W. (Ed). *The Welfare of Pigs* (vol. 11). Martinus Nijhoff, The Hague, pp. 3-13.

McGlone, J. J. and Anderson, D. L. (2002). Synthetic maternal pheromone stimulates feeding behavior and weight gain in weaned pigs. *J. Anim. Sci.* 80(12): 3179–3183.

McGlone, J. J. and Curtis, S. E. (1985). Behavior and performance of weanling pigs in pens equipped with hide areas. *J. Anim. Sci.* 60(1): 20–24.

Meunier-Salaün, M. C., Edwards, S. A. and Robert, S. (2001). Effect of dietary fibre on the behaviour and health of the restricted sow. *Anim. Feed Sci. Technol.* 90(1–2): 53–69.

Nalon, E., Conte, S., Maes, D., Tuyttens, F. A. M. and Devillers, N. (2013). Assessment of lameness and claw lesions in sows. *Livest. Sci.* 156(1–3): 10–23.

Olsson, A. C. and Samuelsson, O. V. (1993). Grouping studies of lactating and newly weaned sows. In: Collins, E. and Boom, C. (Eds). Proceedings of the 4th International Symposium American Society of Agricultural Engineers, Coventry, UK, pp. 475–482.

Olsson, A. C., Svendsen, J. and Reese, D. (1994). Housing of gestating sows in long narrow pens with liquid feeding: function studies and grouping routines in five sow pools. *Swed. J. Agric. Res.* 24: 131–141.

Patterson, J. and Foxcroft, G. (2019). Gilt management for fertility and longevity. *Animals* 9(7): 434.

Patterson, J., Triemert, E., Gustafson, B., Werner, T., Holden, N., Pinilla, J. C. and Foxcroft, G. (2016). Validation of the use of exogenous gonadotropins (PG600) to increase the efficiency of gilt development programs without affecting lifetime productivity in the breeding herd. *J. Anim. Sci.* 94(2): 805–815.

Pedersen, L. J., Heiskanen, T. and Damm, B. I. (2003). Sexual motivation in relation to social rank in pair-housed sows. *Anim. Reprod. Sci.* 75(1–2): 39–53.

Pedersen, L. J., Rojkittikhun, T., Einarsson, S. and Edqvist, L. E. (1993). Postweaning grouped sows: effects of aggression on hormonal patterns and estrus behaviour. *Appl. Anim. Behav. Sci.* 38(1): 25–39.

Pluym, L., Van Nuffel, A., Dewulf, J., Cools, A., Vangroenweghe, F., Van Hoorebeke, S. and Maes, D. (2011). Prevalence and risk factors of claw lesions and lameness in pregnant sows in two types of group housing. *Veterinarni Medicina* 56: 101–109.

Pluym, L., Van Nuffel, A. and Maes, D. (2013). Treatment and prevention of lameness with special emphasis on claw disorders in group-housed sows. *Livest. Sci.* 156(1–3): 36–43.

Potegal, M. (1979). The reinforcing value of several types of aggressive behavior: a review. *Aggr. Behav.* 5(4): 353–373.

Ramonet, Y., Meunier-Salaün, M. C. and Dourmad, J. Y. (1999). High-fiber diets in pregnant sows: digestive utilization and effects on the behavior of the animals. *J. Anim. Sci.* 77(3): 591–599.

Rault, J. L., Morrison, R. S., Hansen, C. F., Hansen, L. U. and Hemsworth, P. H. (2014). Effects of group housing after weaning on sow welfare and sexual behavior. *J. Anim. Sci.* 92(12): 5683–5692.

Remience, V., Wavreille, J., Canart, B. V., Meunier-Salaün, M. C., Prunier, A., Bartiaux-Thill, N., Nicks, B. and Vandenheede, M. (2008). Effects of space allowance on the welfare of dry sows kept in dynamic groups and fed with electronic sow feeder. *Appl. Anim. Behav. Sci.* 112(3–4): 284–296.

Robert, S., Matte, J. J., Farmer, C., Girard, C. L. and Martineau, G. P. (1993). High-fibre diets for sows: effects on stereotypies and adjunctive drinking. *Appl. Anim. Behav. Sci.* 37(4): 297–309.

Robert, S., Rushen, J. and Farmer, C. (1997). Both energy content and bulk of feed affect stereotypic behaviour, heart rate and feeding motivation of female pigs. *Appl. Anim. Behav. Sci.* 54(2-3): 161-171.

Rodenburg, T. B., Bijma, P., Ellen, E. D., Bergsma, R., de Vries, S., Bolhuis, J. E., Kemp, B. and Van Arendonk, J. A. M. (2010). Breeding amiable animals? Improving farm animal welfare by including social effects in breeding programmes. *Anim. Welf.* 19(S): 77-82.

Salak-Johnson, J. L., Niekamp, S. R., Rodriguez-Zas, S. L., Ellis, M. and Curtis, S. E. (2007). Space allowance for dry, pregnant sows in pens: body condition, skin lesions and performance. *J. Anim. Sci.* 85(7): 1758-1769.

Samarakone, T. S. and Gonyou, H. W. (2009). Domestic pigs alter their social strategy in response to social group size. *Appl. Anim. Behav. Sci.* 121(1): 8-15.

Séguin, M. J., Barney, D. and Widowski, T. M. (2006a). Assessment of a group-housing system for gestating sows: effects of space allowance and pen size on the incidence of superficial skin lesions, changes in body condition, and farrowing performance. *J. Swine Health Prod.* 14: 89-96.

Séguin, M. J., Friendship, R. M., Kirkwood, R. N., Zanella, A. J. and Widowski, T. M. (2006b). Effects of boar presence on agonistic behaviour, shoulder scratches, and stress response of bred sows at mixing. *J. Anim. Sci.* 84(5): 1227-1237.

Singleton, W. L. (2001). State of the art in artificial insemination of pigs in the United States. *Theriogenology* 56(8): 1305-1310.

Siswadi, R. and Hughes, P. E. (1995). The efficacy of the boar effect when conducted in a modified detection-mating area (DMA). *Aust. J. Agric. Res.* 46(8): 1517-1523.

Spoolder, H. A. M., Burbidge, J. A., Edwards, S. A., Simmins, P. H. and Lawrence, A. B. (1995). Provision of straw as a foraging substrate reduces the development of excessive chain and bar manipulation in food restricted sows. *Appl. Anim. Behav. Sci.* 43(4): 249-262.

Spoolder, H. A. M., Geudeke, M. J., Van der Peet-Schwering, C. M. C. and Soede, N. M. (2009). Group housing of sows in early pregnancy: a review of success and risk factors. *Livest Sci.* 125: 1-14.

Stevens, B. G., Karlen, G. M., Morrison, R., Gonyou, H. W., Butler, K. L., Kerswell, K. J. and Hemsworth, P. H. (2015). Effects of stage of gestation at mixing on aggression, injuries and stress in sows. *Appl. Anim. Behav. Sci.* 165: 40-46.

Stolba, A. (1988). Ethograms of the domestic pig and European wild boar. *The Libr. Assoc. Study Anim. Behav.*: 287-298.

Stolba, A. and Wood-Gush, D. G. M. (1989). The behaviour of pigs in a semi-natural environment. *Anim. Sci.* 48(2): 419-425.

Strawford, M. L., Li, Y. Z. and Gonyou, H. W. (2008). The effect of management strategies and parity on the behaviour and physiology of gestating sows housed in an electronic sow feeding system. *Can. J. Anim. Sci.* 88(4): 559-567.

Supakorn, C., Stock, J. D., Garay, E., Johnson, A. K. and Stalder, K. J. (2018). Lameness: a principle problem to sow longevity in breeding herds. *CAB Rev.* 13(23): 1-14.

Taylor, I. A., Barnett, J. L. and Cronin, G. M. (1997). Optimum group size for pigs. In: Bottcher, R. W. and Hoff, S. J. (Eds). Proceedings of the 5th International Symposium American Society of Agricultural Engineers, St. Joseph, MI, pp. 965-971.

Telkanranta, H. and Edwards, S. A. (2017). Lifetime consequences of the early physical and social environment of piglets. In: Spinka, M. (Ed.). *Advances in Pig Welfare*. Woodhead Publishing Series in Food Science, Technology and Nutrition, Duxford, United Kingdom, pp. 101-136.

Terlouw, E. M. C. and Lawrence, A. B. (1993). Long-term effects of food allowance and housing on development of stereotypies in pigs. *Appl. Anim. Behav. Sci.* 38(2): 103-126.

Terlouw, E. M. C., Lawrence, A. B. and Illius, A. W. (1991). Influences of feeding level and physical restriction on development of stereotypies in sows. *Anim. Behav.* 42(6): 981-991.

Thornton, K. (1990). *Outdoor Pig Production.* Farming Press Ltd., Ipswich, UK.

Turner, S. P. and Edwards, S. A. (2004). Housing immature domestic pigs in large social groups: implications for social organisation in a hierarchical society. *Appl. Anim. Behav. Sci.* 87(3-4): 239-253.

Turner, S. P., Edwards, S. A. and Bland, V. C. (1999). The influence of drinker allocation and group size on the drinking behaviour, welfare and production of growing pigs. *Anim. Sci.* 68(4): 617-624.

Turner, S. P., Ewen, M., Rooke, J. A. and Edwards, S. A. (2000). The effect of space allowance on performance, aggression and immune competence of growing pigs housed on straw deep-litter at different group sizes. *Livest. Prod. Sci.* 66(1): 47-55.

Turner, S. P., Farnworth, M. J., White, I. M. S., Brotherstone, S., Mendl, M., Knap, P., Penny, P. and Lawrence, A. B. (2006). The accumulation of skin lesions and their use as a predictor of individual aggressiveness in pigs. *Appl. Anim. Behav. Sci.* 96(3-4): 245-259.

Turner, A. I., Hemsworth, P. H. and Tilbrook, A. J. (2005). Susceptibility of reproduction in female pigs to impairment by stress or elevation of cortisol. *Domest. Anim. Endocrinol.* 29(2): 398-410.

Van de Weerd, H. A. and Day, J. E. L. (2009). A review of environmental enrichment for pigs housed in intensive housing systems. *Appl. Anim. Behav. Sci.* 116(1): 1-20.

van Riet, M. M. J., Millet, S., Aluwé, M. and Janssens, G. P. J. (2013). Impact of nutrition on lameness and claw health in sows. *Livest. Sci.* 156(1-3): 24-35.

Velarde, A. (2007). Agonistic behaviour. In: Verlarde, A. and Geers, R. (Eds). *On Farm Monitoring of Pig Welfare.* Wageningen Academic Press, Wageningen, the Netherlands, pp. 53-56.

Verdon, M., Hansen, C. F., Rault, J. L., Jongman, E., Hansen, L. U., Plush, K. and Hemsworth, P. H. (2015a). Effects of group-housing on sow welfare: a review. *J. Anim. Sci.* 93(5): 1999-2017.

Verdon, M., Morrison, R. S. and Hemsworth, P. H. (2015b). Short and long-term repeatability of individual sow aggressiveness. *Anim. Prod. Sci.* 55(12): 1512 (abstract).

Verdon, M., Morrison, R. S., Rice, M. and Hemsworth, P. H. (2016). Individual variation in sow aggressive behaviour and its relationship with sow welfare. *J. Anim. Sci.* 94(3): 1203-1214.

Verdon, M. and Rault, J. -L. (2018). *Aggression in Group Housed Sows and Fattening Pigs.* In: Spinka, M. (Eds.). *Advances in Pig Welfare.* Woodhead Publishing Series in Food Science, Technology and Nutrition, Duxford, United Kingdom, pp. 235-251.

Wang, L. H. and Li, Y. Z. (2016). Effect of continuous access to feeding stalls during mixing on behavior, welfare, and performance of group-housed gestating sows in different social ranks. *Can. J. Anim. Sci.* 96(3): 386-396.

Webb, N. G. and Nilsson, C. (1983). Flooring and injury—an overview. In: Baxter, S. H., Baxter, M. R. and MacCormack, J. A. D. (Eds). *Farm Animal Housing and Welfare.* Martinus Nijhoff, The Hague, The Netherlands, pp. 226-259.

Weng, R. C., Edwards, S. A. and English, P. R. (1998). Behaviour, social interactions and lesion score of group-housed sows in relation to floor space allowance. *Appl. Anim. Behav. Sci.* 59(4): 307–316.

Whay, H. R., Main, D. C. J., Green, L. E. and Webster, A. J. F. (2003). Animal-based measures for the assessment of welfare state of dairy cattle, pigs and laying hens: consensus of expert opinion. *Anim. Welf.* 12: 205–217.

Wiegand, R. M., Gonyou, H. W. and Curtis, S. E. (1994). Pen shape and size: effects on pig behaviour and performance. *Appl. Anim. Behav. Sci.* 39(1): 49–61.

Chapter 3

Optimising sow and piglet welfare during farrowing and lactation

Emma M. Baxter, Animal Behaviour and Welfare Team, Animal and Veterinary Sciences Research Group, SRUC, UK; and Sandra Edwards, Newcastle University, UK

1 Introduction

One of the most contentious and persistent welfare issues in livestock production is the use of housing systems involving confinement. In the pig industry, these include gestation stalls and farrowing crates that together house sows during much of their reproductive life. The welfare detriments to the sow include physical and behavioural restriction leading to physiological and psychological stress. There is growing pressure, expressed through consumer demand, societal opinion and government legislation, to abolish confinement systems such as the farrowing crate (Eurobarometer, 2016). However, farmers have valid concerns regarding such change; they need to achieve good animal performance (i.e. high piglet survival), in systems with acceptable capital, running and labour costs, and which facilitate efficient labour routines and safeguard the operator. Originally, farrowing crates were introduced to achieve these aims, especially to reduce piglet mortality attributed to overlaying by the sow. While reductions in this type of mortality were achieved, piglets also die from other causes (Edwards, 2002) and their mortality is a persistent welfare and economic problem (Baxter and Edwards, 2018). Narrow production-focussed breeding goals exacerbating many of the pre-disposing risk factors for piglet mortality have hindered improvements in survival and further challenge both piglet and sow welfare during farrowing and lactation.

http://dx.doi.org/10.19103/AS.2020.0081.04

There are also welfare concerns for piglets relating to painful husbandry procedures such as tooth resection, tail docking and castration. Furthermore, certain management strategies to rear surplus piglets from large litters pose new risks for piglet welfare. This chapter will highlight some of these welfare challenges, before describing various mitigation strategies to optimise welfare of both sows and piglets in the farrowing environment. Various examples of best practice will be detailed, with a case study describing the development and successful implementation of a high-welfare alternative farrowing and lactation system ('PigSAFE') that attempts to reconcile the 'triangle of needs' relating to the farmer, the sow and her litter.

2 Welfare challenges during farrowing and lactation

The most prominent welfare concerns during farrowing and lactation focus on systems imposing both physical and behavioural restriction on the sow. This is an area of continued scientific and public debate. 'Naturalness' and 'freedom to express natural behaviour' are facets of animal welfare that citizens value highly (Verbeke, 2009; Sørensen and Fraser, 2010; Boogaard et al., 2011) and surveys repeatedly emphasise that sow confinement is in direct opposition to these values (Boogaard et al., 2011; Grunert et al., 2018), with respondents wanting 'spacious farrowing pens with unfixed sows' (Boogaard et al., 2011). Despite cultural differences in attitudes towards animals (D'Silva and Turner, 2012), dissatisfaction with restrictive systems appears to be globally shared. European concerns are well documented (Eurobarometer, 2016) and have resulted in total or partial prohibition of some confinement systems such as sow gestation stalls. Studies in the USA and Canada (Ryan et al., 2015) have shown that North American consumers also oppose the use of stalls and therefore it is likely that they would raise similar objections to 'fixation' of the sow at any period in her life. Brazilian citizens preferred cage-free or free-range systems based on naturalness, animals' freedom to move, and ethics (Yunes et al., 2017). Although animal welfare is a relatively new concept in China, there are indications that the consumers here are also becoming increasingly interested in improving the rearing conditions of farm livestock (You et al., 2014; van de Weerd and Ison, 2019), especially if this impacts upon food safety. Younger consumers, in particular, value environmentally friendly and organic products (de Barcellos et al., 2013; Thøgersen and Zhou, 2012; Chen and Lobo, 2012), suggesting increasing pressure on confinement systems in the future.

While progress has been made worldwide in restricting the use of stalls in gestation (Buller et al., 2018), the predominant maternity system used globally continues to be the farrowing crate. It persists because its principal reason for development persists - high levels of pre-weaning piglet mortality, along with the perception that the majority of these deaths are attributable to overlying by the sow (Edwards, 2002). The farrowing crate reduces crushing mortality by significantly limiting the sow's freedom of movement and provides a safe and cost-effective working environment for farm staff. However, piglets can still die from other causes and the farrowing crate may not be the only solution. Average piglet pre-weaning mortality levels in UK commercial outdoor systems, where the sow experiences zero-confinement throughout farrowing and lactation, are on a par with those in indoor crated systems (12.20% vs. 12.19% respectively), with stillborn mortality actually lower in outdoor units (3.3% vs. 5.8%, AHDB Pork[1]). Furthermore, there is a growing body of evidence demonstrating that the reduction in sow welfare associated with confinement in a farrowing crate actually results in impaired welfare outcomes for her piglets. Confined sows show increased piglet-directed aggression (Jarvis et al., 2006), and greater restlessness during farrowing which increases crushing risk and reduces safe access to the udder (Ocepek and Andersen, 2017). Conversely, in loose-housing environments where sow welfare is improved, piglet advantages include increased maternal carefulness (Ocepek and Andersen, 2017), improved suckling success, as measured by increased IgG levels (Yun et al., 2014), and increased weaning weight (Pedersen et al., 2011; Nowland et al., 2019).

2.1 Piglet welfare challenges

The main welfare challenges for piglets are the pre-disposing factors for mortality. Piglet mortality during farrowing and lactation averages 16-20% per litter (Baxter and Edwards, 2018). Their most vulnerable period is during the first 72h of life, when they can suffer asphyxia, hypothermia, crushing and starvation. These early causes of live-born mortality are often interlinked, and many piglets will become chilled, fail to compete at the udder for vital colostrum and lack energy to move away from the sow when she changes posture (Edwards, 2002). Piglets can also suffer from disease pre-weaning, especially if they fail to suckle colostrum from which they gain passive immunity. Thus, many piglets that die pre-weaning are likely to be exposed to some degree of either pain, hunger and/or fear and stress which could be either acute or chronic. Piglets that are born dead are unlikely to have reached a conscious state and therefore stillbirth is less of a welfare issue for the piglet itself (for a more detailed discussion see Baxter and Edwards, 2018). However, with piglet mortality of approximately 7% attributed to this condition (annual benchmarking figures from commercial herds in USA, UK and DK for 2018, sourced from PigCHAMP, AHDB Pork, SEGES), it is certainly an economic and ethical concern. Many of

1 https://porktools.ahdb.org.uk/prices-stats/costings-herd-performance/ data acquired January 2020.

the pre-disposing risk factors for mortality have been exacerbated by narrow breeding goals focussing on production traits such as hyperprolificacy and lean tissue growth rate, which reduce piglet maturity at birth and increase within-litter competition both pre- and post-natally. A particular concern for piglet welfare is the increased number of piglets born with intrauterine growth retardation (IUGR) (Edwards et al., 2019a). This not only increases neonatal mortality risk, but also has detrimental long-term consequences for physiology and behaviour.

If piglets do survive early life, they will commonly undergo a number of painful husbandry procedures including tooth resection, tail docking and (for males) castration. All of these procedures are performed within the first 7 days of life, usually within 24h post-partum. Piglets may also receive vaccinations and injections of iron, vitamins and antibiotics. While each of these procedures is carried out with some longer-term welfare justification, all cause acute pain and can give rise to additional medium- and long-term welfare detriments.

Piglets can experience further behavioural detriments as a result of barren housing environments, with no access to environmental enrichment or structurally complex surroundings that are known to have positive effects on social and cognitive development (De Jonge et al., 1996; Martin et al., 2015), adaptation to weaning (Oostindjer et al., 2011a), growth rate (Brown et al., 2015; Lawrence et al., 2018), immune responses (van Dixhoorn et al., 2016) and stress regulation mechanisms (Fox et al., 2006). Piglets reared in barren environments often develop poor social skills, display abnormal behaviours both pre- and post-weaning, and lack behavioural flexibility to cope with challenges later in life (for a review see Telkänranta and Edwards, 2018).

2.2 Sow welfare challenges

In conventional pig production, sows are moved to farrowing crates approximately five days before they are due to give birth. A typical farrowing crate has tubular metal bars running horizontally along its length, with additional bars positioned above the sow to prevent escape by jumping or climbing. It measures approximately 2.00m in length and is between 0.45-0.65m wide (Table 4). Flooring is typically fully or partially slatted (plastic or metal) to allow easy removal of waste into slurry pits below. The farrowing crate restricts movements, allowing the sow only enough space to stand up and lie down but not to turn around. The flooring and manure management system generally prohibit the provision of substrate required to help fulfil behavioural needs, such as the performance of highly motivated nest-building behaviour (Wischner et al., 2009). Nest-building is a behavioural pattern typically initiated by sows from 16-24h before they give birth. Its performance reflects a strong evolutionary survival value since, in the wild, these nests protect the young from predators and inclement weather. Nest-building is therefore intrinsically motivated; triggered by endogenous hormones (Algers and Uvnas-Moberg, 2007) and stimulated by extrinsic environmental factors in a sequence including finding a nest-site, rooting the ground, finding, carrying and arranging suitable substrate (Jensen, 1986, 1993). The time sequence in which nest-building behaviour ceases and farrowing starts is influenced by the performance of these activities and the effects that they have on hormone levels that prepare the sow for farrowing. A satisfactory phase of active nest-building occurring 3-7h pre-farrowing coincides with increases in oxytocin level. The sow then becomes less active, lies down and goes into a 'quiet phase' before farrowing begins.

A sow will attempt to perform nest-building behaviour no matter what her environment. Sows kept in farrowing crates or pens with no nest-building material redirect these behaviours to the pen equipment and perform bar-biting, manipulate the drinker and root and paw at the floor (Lawrence et al., 1994). A common misconception is that modern domestic breeds of sow do not nest-build as they have no need, experiencing no risk from predators or inclement weather when housed indoors, and because the behaviour has been bred out of them. This is incorrect and nest-building remains functionally important. Despite years of domestication and selective breeding, and despite protective, warm environments provided to piglets, the modern sow continues to be highly motivated to perform the behavioural patterns of nest-building (Jensen, 2002; Yun and Valros, 2015). When sows are unable to fully perform nest-building due to a lack of space and/or substrate, they show increased stress (Lawrence et al., 1994, 1997; Jarvis et al., 1997, 2001). This adversely affects various aspects of maternal behaviour (for a review see Yun and Valros, 2015) including the progress of farrowing, which has been shown in some (Oliviero et al., 2010), but not all (Jarvis et al., 2001; Hales et al., 2015), studies to increase stillborn mortality. Stress can activate brain opioid pathways, resulting in an inhibition of oxytocin release which is essential to the farrowing process (Lawrence et al., 1992). Opioids play an important role in preventing pain during farrowing (Jarvis et al., 1998; Ison et al., 2018) and may also promote sow passivity towards piglets during the early post-natal period (Jarvis et al., 1999). Disruption of the hormonal control of this passive response may trigger piglet-directed aggression (Ahlström et al., 2002), which is more prevalent when sows are housed in farrowing crates (Jarvis et al., 2004; Ison et al., 2015). Furthermore, reductions in positive maternal behaviours including sustained lateral lying, carefulness when changing posture, including pre-lying sow-piglet communication and responsiveness to piglets are also reported in restrictive systems (Jarvis et al., 1999; Andersen et al., 2014; Yun and Valros, 2015).

Other physiological and physical impacts of confinement on the sow include a reduced ability to thermoregulate (Quiniou and Noblet, 1999; Phillips et al., 2000; Muns et al., 2016), increased risk of hoof, leg and shoulder lesions (Boyle et al., 2002; Leeb et al., 2001) and reduced muscle mass due to prolonged reduction in movement (Barnett et al., 2001). Restrictive housing and the impairment of thermoregulation leads to reduction in feed intake (Pajor, 1998; Muns et al., 2016; Black et al., 1993), which contributes to the development of shoulder lesions/ulcers (Bonde, 2008) and adversely affects milk production and piglet growth rate (Black et al., 1993; Renaudeau and Noblet, 2001). The modern hyperprolific sow is at greater risk of developing injuries when housed in farrowing crates due to her significant increase in size; she is now substantially heavier and longer than her equivalent of 40 years ago (Moustsen et al., 2011). If used as a nurse sow to rear surplus piglets or piglets weaned at an older age, an extended lactation length also results in increased incidence of leg bursae and udder lesions (Ladewig et al., 1984; Sørensen et al., 2016). Research shows that prolonged confinement in a farrowing crate leads to chronic stress in sows (Cronin et al., 1991; Jarvis et al., 2006). This appears to arise from enforced subjection to increasingly persistent piglet attention (Weary et al., 2002) and a lack of ability to express the exploratory and foraging behaviours seen under more extensive conditions as the sow increasingly spends more time away from the nest (Hötzel et al., 2004). Piglets housed in commercial farrowing crates spend more time interacting with their dam, including suckling and udder massage, than piglets reared outdoors (Hötzel et al., 2004). Therefore, it is likely that a lack of exploratory opportunities within conventional farrowing crates also contributes to welfare problems associated with this system.

3 Mitigating welfare challenges

Genetic selection programmes targeting piglet survival traits, including increased robustness and resilience of the neonate, as well as selection for good maternal traits (e.g. udder quality – Balzani et al., 2016; maternal behaviour – Ocepek and Andersen, 2017) would help to tackle some of the welfare detriments described previously. These strategies are discussed elsewhere (e.g. Turner et al., 2018). This chapter will concentrate on optimising nutritional inputs to help combat piglet mortality, as well as describing how changes to the farrowing and lactation environment can improve welfare for both the sow and piglets, while being mindful of practical considerations for the farmers.

3.1 Optimising nutritional inputs

3.1.1 Sow nutrition

There are basic nutritional requirements for sows and piglets that should be met to ensure, at least, provision of the nutrients required for maintenance and production outputs. However, research suggests that the nutritional needs of hyperprolific sows are not always met. Breeding for very large litter sizes imposes more demands on sows, and a need to update nutrient requirements (Strathe et al., 2017). For example, Danish hyperprolific sows have required increases in levels of lysine (by 28%) and crude protein (by 14%) in their rations (Tybirk et al., 2012; Tybirk et al., 2015 – cited in Pedersen et al., 2019). There is also evidence that modern hyperprolific sows have a compromised energy status at the time of farrowing (Feyera et al., 2018), in part due to a prolonged farrowing duration. Farrowing is now reported to last, on average, 7h (Hales et al., 2015), representing an increase of 150 minutes compared to their 'standard' counterparts producing 'normal' sized litters (Olivierio et al., 2010). Protracted farrowing can lead to dystocia, increased stillbirths and piglets born with hypoxia (Björkman et al., 2017; Olivierio et al., 2010), pain for the sow (Mainau et al., 2012; Ison et al., 2016), as well as reduced colostrum yield (Hasan et al., 2019). Given this extended farrowing period, it is likely that sows are suffering from maternal fatigue. Feyera et al. (2018) demonstrated that the time delay between a sow's last meal and the onset of farrowing affected farrowing duration and the number of stillborn piglets. If the time between the two events was below 3h there was no effect on farrowing duration or stillborn mortality, but if the delay was above 6h both were significantly affected. The risk of this occurring is high, given that sows prefer to farrow undisturbed overnight when staff are absent. Feyera et al. (2018) went on to show that the energy status of the sow was depleted as farrowing progressed and recommended that sows were fed three meals a day to reduce the inter-meal interval. They also demonstrated benefits from high-fibre diets, fed from two weeks prior to farrowing and over the transition period, which provided a more sustained energy release after feeding. Feeding high-fibre diets also provides sows with greater gut fill and feelings of satiety (D'Eath et al., 2009), and also relieves constipation (Olivierio et al., 2010), and has benefits for reproductive efficiency when fed post-weaning (for a review, see Jarrett and Ashworth, 2018). Providing sows with pain relief could also mitigate some of the welfare detriments associated with problematic farrowings. Mainau et al. (2012) demonstrated that sows treated with a non-steroidal anti-inflammatory drug (NSAID – meloxicam) spent less time lying inactive on the third day following parturition. They also demonstrated benefits for piglets (increased IgG concentration in serum and enhanced pre-weaning growth) when multiparous sows were given oral meloxicam at the beginning of farrowing (Mainau et al., 2016). Similarly, Homedes et al. (2014) saw a reduction in piglet mortality when sows were given the NSAID ketoprofen.

Other nutritional interventions at the sow level can indirectly influence piglet welfare. These include late gestational feeding of supplements to reduce the impacts of low birth weight on piglet vitality (e.g. palm oil distallate - Amdi et al., 2013a; fatty acids - Rooke et al., 2001; Bontempo and Jiang, 2015;

L-glutamine - Wu et al., 2011). The evidence for improvements in colostrum yield from dietary supplements is conflicting (Theil et al., 2014), but there appears to be greater success in altering colostrum composition via sow nutrition to provide more fat and energy to the piglet (e.g. supplementary leucine metabolite - Nissen et al., 1994; increased dietary fibre – Loisel et al., 2013). Introducing fats and oils into late gestation and lactation diets as a high-energy supplement has been shown to increase sow milk yield, improve neonatal growth and development, and alter sow metabolism (Bontempo and Jiang, 2015). Peng et al. (2019) showed that sows fed a gestation diet supplemented with fat (2% soybean oil) had greater plasma prolactin concentrations, which was thought to be the mechanism benefiting their lactation capabilities. The same study demonstrated increased colostral protein concentration, with others also showing soy oil supplementation in sow lactation diets increases protein concentration in milk (Jones et al., 2002).

3.1.2 Piglets: ensuring optimum colostrum intake

Piglets are completely reliant on their mother for the provision of nutrition in the first few days of life. In order to survive they must quickly get to the udder, acquire a functional teat and suckle colostrum. This not only aids thermoregulation and the acquisition of nutrients and immunoglobulins, but also initiates gut closure, which then reduces the risk of pathogens entering the piglet's systemic circulation (Gaskin and Kelley, 1995). With large litter size, the risk of failing to acquire enough colostrum for survival, growth and development is high (Quesnel et al., 2012; Decaluwé et al., 2014) due to competition at the udder and/or compromised birth weight status (Declerck et al., 2015; Hasan et al., 2019; Amdi et al., 2013b). Piglets need to ingest a minimum of 200g colostrum for survival, but at least 250g is required for normal growth and development (Devillers et al., 2011; Hasan et al., 2019). There is also evidence that, for successful transfer of lymphocytes (B and T cells), cytokines, nucleotides, various growth factors (Bandrick et al., 2011) and other important biochemical signalling factors (Power and Schulkin, 2013), piglets need to suckle colostrum from their own mother. This has implications for optimal farrowing management, and limits the success of early nutritional interventions providing colostrum supplements (e.g. Muns et al., 2014). It is likely that the energy piglets acquire from any supplements has more effect on piglet survival than the provision of immunoglobulins (Thorup et al., 2015; Muns et al., 2015, 2017). This is because the newborn piglet's body reserves do not cover its high-energy needs (e.g. to ensure thermoregulation, locate the udder, compete for a teat and escape potentially dangerous movements of the sow). Providing piglets with extra energy shortly after birth to boost activity could be a means to promote suckling (i.e. the acquisition of sufficient amounts of colostrum) and, consequently, neonatal survival and growth. However, it is difficult to suggest one protocol describing how best to deliver supplements, of which type and to whom, as studies vary in all these factors (e.g. Muns et al., 2014, 2015, 2017; Schmitt et al., 2019a; Englesmann et al., 2019). It is likely that, to be effective, energy supplements should be administered very early in the piglet's life and in not too great a volume in any one meal. This has implications for farrowing management and how best to design a farrowing environment that allows such targeted interventions while still safeguarding sow welfare. Certainly, maximising sow lactational output should be a priority and changes to the farrowing environment to influence colostrum quality (Yun et al., 2014) and provide easy access to the udder (Pedersen et al., 2011) have demonstrated positive effects on suckling success and piglet outcomes.

It has long been established that optimum birth weight is key to piglet survival (Tuchscherer et al., 2000; Edwards, 2002; Baxter et al., 2008) and the long-term prospects for piglets born under that optimum weight are poor because of their impaired suckling success. Roehe (1999) demonstrated that 1.6 kg was optimum, with piglets below this weight suffering a three-fold increase in mortality risk. Twenty years later, selection for hyperprolificacy has increased litter size and reduced average piglet birth weight, and it appears that there has been a shift in this optimum. A multi-study project, measuring birth weight of over 4000 piglets from 394 litters at four different commercial farms, concluded that piglets born below 1.11 kg suffered a six-fold increase in mortality risk (Feldpausch et al., 2019). The shift in average birth weight accompanying hyperprolificacy has not resulted in small but robust piglets more capable of survival. It is more likely that increased husbandry interventions for large litters have influenced the survival prospects of these smaller piglets. This is supported by recent evidence suggesting that, for piglets born under 1.8 kg in weight, it is their weaning weight that can best predict their life-time growth rate (Douglas et al., 2013), indicating that such piglets can display catch-up growth if provided with the correct resources. Therefore, ensuring high milk intake during the pre-weaning period for smaller piglets may help them to reach their longer-term growth potential. Supplementary milk provision in the farrowing pen, in addition to a stable suckling environment, may aid this endeavour. However, it should be noted that these interventions concentrate on growth and survival. Long-term health and welfare implications for piglets born into large litters and with low birth weight are yet to be fully understood (Edwards et al., 2019a).

3.1.3 Large litter management and ensuring optimum milk intake

If piglets are born into a very large litter and there are more piglets than functional teats, spilt suckling can be performed to ensure all piglets get as much colostrum as possible from their own mother (Baxter et al., 2013). This involves splitting the litter into two groups, usually based on their weight and/or vitality (Kyriazakis and Edwards, 1986; Donovan and Dritz, 2000). The lightest/weakest piglets are allowed access to the udder first, while the heaviest/strongest ones are enclosed in a heated creep area or a designated box. The piglet groups are then alternated after a few successful sucklings and this continues until fostering opportunities present themselves.

Cross-fostering can ensure a more even number and weight distribution of piglets on the udder to reduce competition and, if performed correctly in the first 48h of life (after a recommended 6h minimum with their own mother), it can be successful in promoting greater survival rates and growth of low birthweight piglets (Alexopoulos et al., 2018; Douglas et al., 2014). However, if performed too early (before 6h post-farrowing), too late (after 48h post-farrowing - Price et al., 1994; Straw et al., 1998; Horrell and Bennett, 1981) or over-performed it can be disruptive, stressful and counterproductive (Robert and Martineau, 2001; Straw et al., 1998). Piglets develop a teat order within the first few days of life, where they become faithful to an individual teat and will vigorously defend it (Puppe and Tuchscherer, 1999). Teat order stability comes from cohesive group suckling behaviour (Skok and Škorjanc, 2014) which promotes growth and development of piglets. Cross-fostering can disrupt this stability, and the introduction of new piglets into a litter that has already established its teat order can lead to fights causing facial injuries to piglets, sow udder injuries and disrupted milk ejections, which in turn can impact growth and development of both fostered and resident piglets (for a review, see Baxter et al., 2013).

Simple cross-fostering cannot solve the problem of supernumerary piglets in a whole farrowing batch, which is now commonplace in hyperprolific herds. More complex strategies need to be adopted to deal with these extra piglets throughout lactation. These include nurse sow strategies (i.e. use of sows that have had their own piglets early weaned in order to free up spare teats to rear surplus piglets) or artificial rearing of surplus piglets. Both practices are reviewed in detail in Baxter et al. (2013, 2018) and have implications for both piglet and sow welfare (Sørensen et al., 2016; Schmitt et al., 2019b). For nurse sow strategies, some studies (Kobek-Kjeldager et al., 2020) but not all (Schmitt et al., 2019c) show detrimental impacts on piglet weaning weight. Schmitt et al. (2019c) did find significant weight reduction in the first week after fostering when comparing piglets moved to nurse sows with piglets left with their mother, but by weaning the differences were no longer significant. They selected the heaviest and most robust piglets for fostering and chose nurse sows with good temperament and udder quality, which are likely to be factors in how successful these strategies can be. It should be noted that one major threat to piglet health is introduced by use of nurse sows; that is, the compromise of biosecurity by breaking batch integrity in the farrowing system (Calderón Díaz et al., 2017).

Minimising exposure of suckling piglets to pathogens is an integral part of controlling pre-weaning mortality and 'all-in-all-out' (AIAO) management of farrowing batches is the key to this. As nurse sow strategies involve some form of early weaning (at least relative to EU regulations), there are also implications for piglet welfare because of the known stressors involved in early separation from the dam (Drake et al., 2008).

It can be argued that artificial rearing is worse for piglet welfare than nurse sow strategies, since it typically involves very early separation from the mother and removal to a more barren environment. Here piglets are placed in specialised enclosures, usually located in a separate room or sitting above the farrowing crate, where they will be fed milk replacer until weaning age (usually at 28 days) (Schmitt et al., 2019d). The enclosures also contain a heat lamp to ensure thermal comfort of the piglets, milk and water cups that can be activated by nudging with their snout, and solid 'creep' food. The fact that piglets are fed ad libitum in a controlled environment, where the risk of crushing is removed, is quite attractive to farmers who may not be able to implement nurse sow strategies. Additionally, some studies have shown artificially reared piglets to have higher weaning weights (van Beirendonck et al., 2015; Cabrera et al., 2010). However, others have observed short-term (De Vos et al., 2014) or sustained (Schmitt et al., 2019d) impairments of growth in artificially-reared piglets as well as behavioural abnormality.

3.1.4 Pre-weaning solid feed intake

Piglets kept under natural or semi-natural conditions begin to forage for solid feed from about two weeks of age and gradually increase their intake as the sow's milk production declines. Introducing piglets to solid feed pre-weaning can reduce weaning stress, which can otherwise contribute to compromised gut and immune function, reduced feed-intake, growth rate, health and welfare (Weary et al., 2008). While early introduction of a starter diet is typically practiced on commercial farms from the second week of life, further optimisation could consider the palatability and presentation of the feed provided. In a series of experiments by Oostindjer et al. (reviewed in Oostindjer et al., 2014) it was demonstrated that post-weaning feed intake was improved when piglets could eat together with their mothers pre-weaning, and that flavour learning was also effective in promoting growth and reducing weaning stress. In these studies, the authors designed a shared feeding station allowing piglets to eat with their mother (Oostindjer et al., 2011b) and gave sows flavoured gestational feed before presenting piglets with the same flavour in both the feed and the air post-weaning (Oostindjer et al., 2011c). Middelkoop et al. (2018) also studied the impact of altering food presentation and flavour on piglet feed intake pre- and post-weaning. As early as two days post-farrowing, one group of piglets was presented with a choice of two diets differing in production method, size, flavour, ingredient composition and nutrient profile, smell, texture and colour (i.e. dietary diversity – DD). The other group had feed that was changed in flavour every six days (i.e. flavour novelty – FN). The DD piglets were more interested in feed and had higher intake than the FN group and it was thought that the diversity stimulated feed exploration and therefore intake. The same authors also demonstrated that the type of creep feeder influenced feed intake (Middelkoop et al., 2019). A round conventional feeder was less attractive to the piglets than a play-feeder, whereby the same round design was adorned with enrichment materials to stimulate rooting and exploratory behaviour.

Providing an enriched neonatal environment can also be beneficial for pre-weaning growth rate. Oostindjer et al. (2010) showed that piglets developed feeding habits better in a loose-housed farrowing and lactation environment compared to a conventional crate system. This is supported by evidence that an enriched neonatal environment that stimulates play behaviour can improve growth of litters (Brown et al., 2015). A combination of nutritional and environmental enrichment may be particularly valuable in mitigating the effects of being born with low birth weight. Understanding the importance of environmental enrichment is one aspect of a larger discussion about providing for the biological needs of the animals to optimise their welfare and enhance performance.

3.2 Biological specifications for optimal housing design

Although the definition of animal welfare has historically separated the animal's physical health and well-being from its psychological health (i.e. the 'biological functioning' vs. the 'feelings' schools of thought – Duncan, 2005), providing for both is imperative for good animal welfare. If a 'need' is denied, this results in a negative welfare state (Jensen and Toates, 1993). Where a need is life-sustaining (e.g. food and water), prolonged failure to make provision for it would be fatal. Therefore, hierarchies of need have been proposed (Duncan, 2005) with these basic life-sustaining needs understandably prioritised. Within livestock production, behavioural needs are often demoted or ignored within the hierarchy, perhaps because of a failure to recognise (or for animal welfare scientists to communicate) their importance for biological functioning. This may actually be counter-productive for animal productivity, as it is well established that performance of species-typical behaviour contributes to an animal's biological fitness (Hamilton, 1964a,b). Behaviours (e.g. nest-building) that have persisted despite centuries of domestication and selection pressure for production traits are those that remain biologically significant.

Thus, biological needs should include physiological, physical and behavioural necessities, and providing for these should promote welfare, health and performance. Meeting biological needs in the farrowing and lactation environment requires the identification of what is most important for both piglets and sows, and then determining the best way to make provision. When designing a new housing system, this means understanding that the animals themselves are active participants in the process and not just passive receivers of the husbandry systems they are put into, as typically practiced since the industrialisation of farming (Bos and Koerkamp, 2009). Many of the health, welfare and environmental problems facing livestock sectors today arise from the dominant post-war policies of modernisation, which stem, understandably, from the focus on ensuring domestic food security at an affordable price. Re-designing the environment with the occupants in mind should redress the balance.

In the case of farrowing and lactation, Baxter et al. (2011a) attempted this approach by first describing each physiological and behavioural need, as indicated by sound biological evidence, and then determining which of these needs are sensitive to the physical environment. They subsequently translated this information into design criteria for a maternity unit built to the biological specifications of the pigs (as summarised in Table 1). This initial task formed the basis for re-designing the farrowing and lactation environment to maximise welfare (The PigSAFE project – see Section 3.4).

3.2.1 Space

The sow needs space for finding and creating a nest, giving birth (parturition), suckling and interacting with her piglets, feeding, drinking, defaecation as well as space for re-integration with her social group. The piglet's spatial needs include space in which to be born, search for the udder, find a teat and suckle. It needs space to rest and keep warm, interact with its siblings and mother, as well as to avoid her during posture changes, and to grow and develop. With current genetics, this means providing space for at least 14 piglets (or the likely maximum number of piglets a sow could support) until weaning at 28 days post-partum.

Quantifying the spatial needs of the animals involves measuring the size of the modern-day domestic sow and her piglets (at weaning) based on the 95th percentile (i.e. largest) individuals. Then, determining both the static and dynamic space requirements for different postures and posture changes, turning around, piglet gathering, separation of functional areas, etc. (Baxter and Schwaller, 1983; Petherick, 1983, Gonyou et al., 2006). Allometry, based on the equation $A = k*W^{0.667}$ (where A=area, k= a constant and W= live-weight), can be used to estimate the space an animal occupies as a consequence of its mass (Baxter and Schwaller, 1983). This equation is very general and assumes all animals are the same shape and that this is consistent over time. In general,

Table 1 Summary of sows' and piglets' biological specifications for a farrowing and lactation system and estimated 'values' required to meet their needs. More detailed descriptions are given in Baxter et al. (2011a, 2018) and any new information published since these manuscripts has been incorporated. Where no new information has materialised and values for components cannot be given, 'further research' is noted

Component of system	Sows	Value of required specification at minimum level to meet needs	Piglets	Value of required specification at minimum to ideal level
Space	Increased activity for nest-site seeking	4.9m²	Parturition	2.79m²
	Hygiene – separate dunging space from feeding and lying area	Separate dunging area from nest and feed sites.	Udder access for suckling throughout lactation	2.79m²
	Feeding and foraging	Separate feeding area from nest and dung sites	Protection, safe lying area for parturition and nest-occupation	Separate space inaccessible to the sow e.g. 0.8m² per 10–12 neonates
	Turn-around nest space for piglet inspection and gathering behaviour	Floor space = 2.44m², planar space = 3.17m². Further research needed		
	Lateral lying and parturition	2.79m²	Protected lying area during lactation	0.96m² for 14, four-week-old piglets
	Thermal comfort via posture changes	2.44m²	Area for feed trough to introduce starter diet and area for supplying supplementary nutrition/energy (separate from the sow)	Provide in the creep, interacts with above
	Nest-departure	Separate area from nest site		
	Gradual separation from piglets and sow-controlled nursing	Separate space inaccessible to piglets		
	Social contact with other sows	Separate space inaccessible to piglets to allow contact between sows, but if sows fully reintegrated before weaning larger space to allow body language assessment and/or fighting to establish dominance – further research needed to determine minimum space per sow.	Hygiene	Separate area for dunging, interacts with flooring

Substrate	Nest-building – carrying and manipulating	2 kg long-stemmed straw	Foraging, nutritional development	Earth-like materials (e.g. peat) Further research needed on quantity. Novelty requires fresh input daily. Complex materials (e.g. branches) preferred.
	Complete nest	2 kg long-stemmed straw and branches	Enrichment, social and cognitive development	2.5cm of straw, interacts with flooring
	Udder comfort	Further research needed, interaction with floor properties	Thermal comfort during parturition	Further research needed, interacts with thermal comfort and flooring properties
	Thermal comfort during nest-building and parturition	2 kg long-stemmed straw in nest site – interaction with flooring and room temperature.	Physical comfort	Deep bedding – 10-12cm, interacts with flooring
	Foraging material	Further research needed	Protection	
Walls	Enclosure/Isolation of nest	3 solid-sided walls (cul-de-sac)	Protection from sow posture changes	Sloped wall or protection bars
	Visual and physical contact with non-litter pigs	Vertical barred area with void wide enough to allow at least nasal contact between pigs	Social contact (visual and physical)	Vertical barred area
		Solid at base with separation between pens	Hygiene	Solid walls (at least at bottom of penning) separating other litters
	Supported posture changes	Solid sloped or vertical walls	Thermal comfort	Solid walls with thermal resistance properties to limit heat loss via radiation – interacts with substrate and flooring
	Lack of disturbance	Further research needed		

(Continued)

Table 1 (*Continued*)

Component of system	Sows	Value of required specification at minimum level to meet needs	Piglets	Value of required specification at minimum to ideal level
Flooring	Nest-building - digging, rooting and hollowing	Malleable (e.g. earthen) or solid to accommodate deep substrate	Thermal comfort during parturition and first 24h of life	High thermal resistance - e.g. rubber matting or deep substrate or under-floor/localised heating
	Nest-building and parturition	Solid to accommodate substrate	Thermal comfort during lactation	High thermal resistance - e.g. rubber matting or deep substrate (see above) or under-floor/localised heating
	Thermal comfort during nest-building, parturition and lactation	Temperature differentials in separate areas allowing choice. High thermal resistance e.g. rubber matting or deep substrate. Low thermal resistance e.g. metal.	Physical comfort - avoiding suckling injury, promoting suckling behaviour	Solid flooring with minimal abrasiveness and well-maintained (e.g. rubber matting or specialised screed with non-slip properties), interacts with substrate
	Physical comfort - avoiding injury, promoting suckling behaviour	Non-slip surface e.g. rubber matting or plastic-coated metal. Minimal abrasiveness (interacts with substrate). Solid to avoid teat injuries. Slatted area	Protection from fatal crushing by the sow	Slatted flooring with void width no more than 10mm and rounded edges. Interacts with temperature (see general)
	Hygiene	Gradation of floor with slope away from lying area. E.g. 10% slope for drainage	Hygiene	

General			
Thermal comfort	Ambient temperature 12–22°C, interactions with substrate and flooring	Health treatment for injuries, vaccines, etc.	Safe area for handling required, interacts with space
High feed intake	See space and thermal comfort	Promote weaning, reduce nutritional stress and encourage increased feed and water intake	Suitable solid food, inaccessible to the sow – interacts with space and substrate. Provide feed tray and sufficient space to allow social facilitation
Low light in nest-site		Thermal comfort	Localised heat source set at thermo-neutral temperature (e.g. 34°C at birth) – interacts with substrate
		Hygiene	Temperature differentials to encourage dunging outside of nest site – interacts with flooring

because these equations err on the side of caution, they are still relevant today for calculating static space requirements (Gonyou et al., 2006). However, many systems are so fixed in their pen dimensions that they have provided no flexibility for coping with changes in the animal's size and productivity over time, or changes to husbandry practices that impact on space. The farrowing crate is such a system. Because the modern-day sow is substantially larger than her counterparts of the 1980s (Moustsen et al., 2011) and she rears a larger litter for a longer lactation length, many farrowing pen designs do not allow sufficient space for hyperprolific litters (Pedersen et al., 2013). Piglets can neither rest together in a thermally comfortable area nor have unobstructed suckling. EU Council Directive 2008/120/EC states, 'When a farrowing crate is used, the piglets must have sufficient space to be able to suckle without difficulties'. The length of a four-week-old piglet (approximately 0.5 m – Moustsen and Poulsen, 2004), therefore needs to be provided at each side of the crate to avoid fighting caused by blocking of piglets' access to preferred teats. Suckling success is promoted when such constraints are not in place, with greater milk let-down reported when sows show better udder access in loose housing (Pedersen et al., 2011). The detailed body measurements provided by Moustsen and colleagues allow calculations for the necessary creep dimensions, suckling space and space for fixtures and fittings within a pen to prevent injury (e.g. creep bar widths). These data can then be applied to determine the detailed design dimensions for any new farrowing and lactation system (Table 1).

3.2.2 Substrate

Although there is strong evidence that space is more important than substrate for allowing the behavioural expression of nest-building (Jarvis et al., 2002; Cronin et al., 1994; Hartsock and Barczewski, 1997), substrate is still very important for sows and piglets. Given that the majority of commercial sows are housed in conventional crated environments (>90% in selected European countries, >85% in Australia and New Zealand – Baxter and Edwards, 2016[2]), it is important to investigate more fully which substrates are of relevance to the sow before and after farrowing in order to improve her welfare, maternal behaviour and the welfare of her piglets. In the periparturient period, the sow needs to collect and arrange suitable substrate for nest-building, and to achieve thermal and physical comfort for herself and her piglets. Substrate also facilitates foraging and nutritional development, providing enrichment which is important for the social and cognitive development of piglets.

2 Proceedings of Free Farrowing Workshop 2016, Belfast NI (eds, Baxter, E. M. and Edwards, S. A.). Proceedings available at https://www.freefarrowing.org/info/2/research/45/free_farrowing_workshops.

The type of substrate provided should have properties which satisfy the biological needs. For nest-building, the material needs to be suitable to provide feedback to the sow signalling that the nest is complete. Without this, some sows may continue to be motivated to nest-build even during farrowing (Thodberg et al., 1999; Damm et al., 2000), which constitutes a risk for the newborn piglets. Substrate manipulation to allow arrangement, as well as a sense of enclosure, appears to give a sense of completion and, in preference tests, sows chose pre-formed nest-sites that also offered a sufficient quantity of straw to satisfy nest-building behaviour (Arey, 1992). 'Sufficient quantity' is a rather open description but it is one found in the European legislation outlining provision of enrichment material at all stages of production. Council Directive 2008/120/EC states that 'pigs must have permanent access to a sufficient quantity of material to enable proper investigation and manipulation activities'. For nest-building, studies have suggested as little as 2 kg of long-stemmed straw to be sufficient (on solid floors – L. J. Pedersen *personal communication*), whereas voluntary use of as much as 255 kg of mixed substrate has been reported (outdoor semi-natural systems - Zanella and Zanella, 1993). Practicality is important and interactions with the flooring and waste management system are factors requiring consideration. EU legislation says that all sows must have material for nest-building but provide possibilities for derogation, stating that 'unless the slurry system makes provision unfeasible'. Straw is widely considered as the gold-standard material for its manipulable as well as thermal properties (Rosvold et al., 2018; Mount, 1967; Wathes and Whittemore, 2006). Westin et al. (2014) demonstrated that, if used strategically, a slurry system could cope with large quantities of straw at the time of nest-building. They provided 15–20 kg of long-stemmed straw when sows moved in to the farrowing accommodation and did not subsequently remove (unless soiled) or add to this, allowing the straw to gradually fall through the slatted part of the pen and be removed with the excreta. Alternative nesting materials have also been studied (e.g. cloth tassels - Widowski and Curtis, 1990; burlap sacks – Bolhuis et al., 2018; Plush et al., 2019; lucerne hay - Edwards et al., 2019b; peat - Rosvold and Andersen, 2019; newspaper, sawdust, shredded paper - Swan et al., 2018). Something as simple as a cloth tassel tied to pen fittings that animals can pull, tear and manipulate may provide welfare benefits, even though the material cannot result in the building of a suitable nest. This is particularly relevant for animals kept in crates, as most of the tassel remains attached to the front of the crate where sows can continue to access it. In contrast, substrate on the floor often gets pushed to the rear of the crate or out of reach during nest-building, with potentially frustrating consequences. Swan et al. (2018) specifically looked at nest-building in crates and compared six materials, either attached to the farrowing crate or in amounts of 1–2 L placed in front of the sow. These included point-source objects offered on the ground or to the side, wood shavings, straw, shredded paper and whole newspaper. Sows showed nest-building behaviour with all materials, but a functionality assessment saw straw, wood shavings and newspaper tested further. There were different benefits for the selected options: the newspaper group performed more nest-building and fewer bar-biting activities, while piglet mortality during the entire lactation period was lower in the straw group than the other groups. This study demonstrates benefits to the piglets of maternal nest-building and is supported by studies, both in crates and loose-housing, that show providing for nest-building can improve suckling success and growth rate in piglets (e.g. crates - Edwards et al., 2019b; loose housing - Yun et al., 2014; Pedersen et al., 2011; Plush et al., 2019) and can reduce stillbirths (e.g. Rosvold and Andersen, 2019; Edwards et al., 2019b). Edwards et al. (2019b) also showed that sows in crates continued to interact with provided substrate (in the form of lucerne hay) throughout lactation, suggesting they still found the enrichment rewarding after the nest-building phase.

Enriching the environment throughout lactation will also benefit the piglets. Studies have shown that providing substrates, point-source objects, opportunities for play and socialisation improves cognitive and social skills (Martin et al., 2015), growth rate (Brown et al., 2015), gut health (Oostindjer et al., 2010) and immunity (van Dixhoorn et al., 2016), and can reduce weaning stress. Which material and how much to provide to best achieve these benefits is still difficult to quantify but, as a general rule of thumb, it appears that diversity of the environment will accrue the most benefits to different aspects of the piglets' welfare. It is fast becoming recognised that an optimised gut microbiota can enhance disease resistance (Patil et al., 2020) and, as efforts are being made to reduce antibiotic use, finding alternative methods to achieve such outcomes are being sought. Increasing diet complexity, by introducing foraging opportunities to piglets pre-weaning, can increase microbial diversity and the presence of beneficial microbes, reducing the risk of post-weaning diarrhoea. Reduction of stress is also important for healthy gut development, with early-life stressful experiences demonstrated, in rodents, to lead to dysfunction of the intestinal barrier (O'Mahony et al., 2009). This is relevant for piglets, which may experience a number of acutely stressful husbandry procedures within the first few days of life.

3.2.3 Walls

The benefits of suitable walls in a farrowing environment include imparting a sense of enclosure for nest-building, providing a solid surface to support sow posture changes and thereby providing protection for piglets and also providing opportunities for both seclusion from conspecifics and interactions with them. When sows choose nest-sites they prefer a 'cul-de-sac' arrangement, providing protection on three sides, and a nest-entrance with a view, presumably for vigilance against potential threats. The degree of enclosure afforded by the nest-site will influence disturbance, and disturbance is known to adversely affect oxytocin levels (Lawrence et al., 1992). Providing sows with a supportive surface to lean against during their descent when lying reduces crushing (Baxter, 1991, Marchant et al., 2001; Damm et al., 2006) and sows prefer sloped or vertical walls to lie down against rather than farrowing rails (Damm et al., 2006). Manufacturers of outdoor farrowing huts or arks recognise that having angled walls is protective for piglets, as they prevent the sow from fully contacting the lower walls when changing posture and therefore trapping piglets. Recreating these protective elements in indoor environments requires quantification of the best dimensions/angles for the sloped walls to provide piglet escape zones, while being an attractive prospect to encourage supported lying behaviour by sows. Moustsen (2006 – cited in Pedersen et al., 2013) provided these details and alternative systems have successfully incorporated such features (Baxter et al., 2015).

While enclosure is important in the periparturient period, as lactation progresses nest-departure and re-integration of the sow and litter into the social group occur in natural conditions when the piglets are about two weeks old. Sows leave the nest to increase their foraging territory and increase feed intake during the metabolically demanding lactation phase. Piglets also benefit from this foraging activity. At the time of reintegration, prior familiarity with other group members minimises aggressive interactions that might impair fitness. Under natural or semi-natural conditions, pigs maintain established groups where aggression is regulated via an 'avoidance order', with specific behavioural patterns reducing risk of attacks by dominant individuals (Jensen, 2002). When a sow leaves the group to farrow, the longer she remains isolated, the more challenging it is for her and her litter to re-integrate. Early socialisation with other litters during lactation benefits piglets by reducing the impact of weaning (Pajor et al., 1999; Hessel et al., 2006), particularly the effects of mixing aggression, and improving piglet social skills post-weaning (Morgan et al., 2014). Alternative farrowing systems which include group housing or multisuckling allow this more natural reintegration; this was popular in the early 1980s, especially in Sweden. They are a cheaper alternative to individual housing throughout lactation and, if the benefits outlined above can be realised, they seem like a sensible choice. However, in general, group housing systems have returned poor production figures for piglet mortality (see Baxter et al., 2012 for a review). More recent research in Germany (Bohnenkamp et al., 2013), the Netherlands (van Nieuwamerongen et al., 2015) and Australia (Verdon et al., 2019; Greenwood et al., 2019) has revisited these systems, partly as a way to increase lactation length and reduce weaning stress, and thus reduce the post-weaning reliance on antibiotic interventions to maintain piglet health. However very careful management is required to ensure success of such systems (van Nieuwamerongen et al., 2014; Thomsson et al., 2016). Facilitating some form of social contact between different sows and litters in individual housing systems involves designing partitions with areas to allow fence-line contact between neighbours. Although this does not allow for full reintegration before weaning, it has been shown to benefit post-weaning outcomes such as reducing aggression upon mixing (Martin et al., 2015). Whether walls are solid or barred will also impact on dunging behaviour (Moustsen and Jensen, 2008) and the microclimate within the pen. Providing cooler (better ventilated) areas will reduce the risk of heat stress, particularly for sows during lactation (Muns et al., 2016), although thermal comfort also depends on substrate provision and floor type.

3.2.4 Flooring

Suitable flooring is one of the topics that requires significant attention in pig production systems, especially when it comes to providing for biological needs. Flooring qualities for locomotion and lying behaviour are based on many different aspects: friction, abrasiveness, hardness, surface profile and thermal properties (Lensink et al., 2013). For all livestock, floors should provide physical and thermal comfort when lying, should not lead to injury or slipping when standing and walking, but should also not be too abrasive. To promote good hygiene, they should facilitate easy cleaning, which will reduce transmission of infectious diseases (Lensink et al., 2013). There are consequently many trade-offs to consider; deciding on the optimal flooring in the farrowing and lactation environment is particularly complicated because of the diverse needs of the sows and piglets at different stages (i.e. nest-building, parturition, lactation) and because flooring interacts with many other design specifications when trying to meet those needs. For nest-building, a sow would ideally want a malleable floor to dig a hollow by rooting the ground. While this is achievable with soil under natural conditions, a proxy indoors involves provision of plentiful substrate, which means that the flooring needs to be solid to retain this substrate. This is also the case for substrate provided as enrichment material and to improve comfort by reducing the risk of development of shoulder ulcers in sows during prolonged lying bouts (Rolandsdotter et al., 2009; Rioja-Lang et al., 2018) and leg injuries in piglets during suckling (Zoric et al., 2009; Lewis et al., 2005; Mouttotou and Green, 1999). However, excreta build-up on solid floors can increase the risk of disease, and self-cleaning slatted systems tend to promote good pen hygiene (as well as reducing labour) (Rantzer and Svendsen, 2001). Good hygiene can be further promoted by clear differentiation between lying areas and areas designed for defaecation. Providing part-slatted flooring in pens helps to designate these areas, and also to provide temperature differentials as slatted floors have different thermal properties to solid floors. If given the correct cues, pigs will actively separate their dunging, resting and feeding sites (Randall et al., 1983). They will also seek corners in which to defecate (Wiegand et al., 1994), which has implications for pen design.

If plentiful substrate is not provided, soft surfaces, such as rubber matting, may allow some deformation of the surface. This reduces contact pressure and mechanical stress on the body, and can aid healing if sows have shoulder sores (Zurbrigg, 2006). Rubber mats can also reduce carpal lesions of the piglets (Courboulay et al., 2000). However, the quality and abrasiveness of the mat are important to prevent excessive wear and tear, as well as to prevent slipping. Some studies suggest rubber flooring is less slippery (Boylet et al., 2000), but others report a 'film of slurry' when mats were used in farrowing houses (Calderón Díaz et al., 2013) which will increase the slip risk for sows and impair hygiene. Flooring is a very difficult system component to perfect and, when designing a new indoor system, providing different floor types in different functional areas within the same pen is likely to be the best way to meet all biological needs.

3.2.5 Thermal comfort

The thermo-neutral zones of sows and newborn piglets are markedly different. A sow's evaporative critical temperature is the upper limit of the thermo-neutral zone and represents the temperature at which evaporative heat loss begins to increase and heat stress develops. This depends on a variety of factors associated with the sow's ability to thermoregulate and is heavily influenced by components of the housing system and ambient temperature. It has been postulated that peri-parturient sows will actively choose a nest-site based on the most thermally resistant flooring (Pedersen et al., 2006). In a study by Hunt and Petchey (1987), sows given the choice of either a concrete floor or a rubber floor with varying amounts of straw chose the mat with the greatest amount of straw. However this choice might have been dictated by comfort or substrate availability, as subsequent studies investigating nest-site choice have shown no difference between heated or room-temperature flooring with the same properties and substrate availability (Pedersen et al., 2006; Baxter et al., 2015). It is generally thought that peri-parturient sows can tolerate slightly higher temperatures than their typical thermal comfort zone during lactation of 10-20°C (Black et al., 1993) and it could be postulated that this tolerance enhances biological fitness as piglets are extremely cold sensitive. This tolerance may be limited, however, as Muns et al. (2016) found that sows in farrowing crates experiencing 25°C around parturition altered their postural behaviour relative to those kept at 20°C. They reacted to the thermal challenge with higher respiration rate, but both their rectal and udder temperatures were elevated, indicating that they were not able to compensate for the higher ambient temperature. When sows are loose-housed they have more possibility to regulate their temperature by altering their postures and choosing different areas in their environment. A study investigating three room temperatures (15°C, 20°C and 25°C) for lactating sows kept loose in pens with a partly slatted concrete floor showed that sows used the cooler slatted floor for behavioural thermoregulation by resting in this zone in between daily activity bouts (Malmkvist et al., 2012). During lactation, sows are more at risk from heat stress due to the higher metabolic activity associated with elevated feed intake and milk synthesis (Williams et al., 2013). Sows unable to thermoregulate will experience hyperthermia/heat stress, which can impact milk production and therefore piglet growth rate and survival. Thus, loose housing with different thermal zones may increase sow thermal comfort and positively affect piglet growth (e.g. Pedersen et al., 2011; Oostindjer et al., 2010).

The thermal challenge for piglets is avoiding hypothermia. At the time of birth they experience a sudden drop in ambient temperature (of approximately 15–20°C – Herpin et al., 2002). Consequently, the lower limit of the thermoneutral zone (the lower critical temperature – approximately 34°C Mount, 1968), is exceeded, resulting in chilling. It is impossible to keep the piglets in their thermoneutral zone by raising room temperature, as heat stress for the sow could be fatal. Thus, providing a suitable microclimate in a designated piglet zone is a necessity, and substrate, flooring and the degree of enclosure in the nest-site all influence the effectiveness of this. Deep bedding slows heat loss, having a thermal resistance 11 times greater than that of concrete slats and 22 times greater than solid, wet concrete flooring (Wathes and Whittemore, 2006). Mount (1967) demonstrated that piglets in contact with a concrete floor lost 40% more heat than those in contact with 2.5cm of straw. This is, in part, why straw management is such an important factor in outdoor pig production. In the absence of deep bedding, an artificially heated creep area of suitable size (Table 1) can provide an effective microclimate, provided that piglets can be attracted to use this area at an early age.

3.2.6 General system components

Although there are some advances to be made in applying nutritional strategies (see Section 3.1), it is assumed that the fundamental physiological requirements for food and water are generally well catered for. However, when sows are loose-housed, placement of feeders and drinkers can influence the spatial organisation of behaviour, especially excretory behaviour (Moustsen and Jensen, 2008; Andersen and Pedersen, 2011; Ocepek et al., 2018), and could influence farrowing location (Baxter et al., 2015). Sows will typically eat and then turn away from their feeding area to defecate, often orientating themselves so that their head faces away from their feed site (Moustsen et al., 2007; Moustsen and Jensen, 2008; Andersen and Pedersen, 2011). Incorporating barred partitions between pens in the designated dunging area, and a solid dividing wall between designated dunging and nesting sites, can improve pen hygiene (Moustsen and Jensen, 2008). However, there is large individual variation in dunging patterns in loose-housed sows (Bøe et al., 2016) and maintaining hygiene continues to be a subject needing research activity (Hansen, 2018). This is not just because good pen hygiene reduces the risk of disease transmission, but it would also encourage greater acceptance of loose-housing systems. Cleaning out farrowing pens and providing new bedding was estimated to require 33% of total daily work time for Swedish farmers on 35 commercial pig farms (Mattsson et al., 2004). The stockperson is a key contributor to animal welfare and should be considered as an important participant in system design. Stockperson intervention to maximise piglet survival has increased in the wake of superprolific breeding programs, where the number of piglets born regularly surpasses the sow's ability to rear them. Large litter size management involves handling of piglets and sows, cross-fostering, split-suckling, establishment of nurse sows, and provision of supplementary nutrition (Baxter et al., 2013, 2018). This represents a challenge in any system but could be regarded as more challenging in loose-housing conditions (Rosvold et al., 2017). Facilitating stockperson interventions while safeguarding the welfare of the sow is therefore an important element of system design. Several Danish studies have evaluated loose-housing systems and pinpoint large litter size as the main risk factor in allowing sows to be loose all the time (i.e. zero-confinement) (e.g. Hales et al., 2015). As such, there has been sustained interest in developing 'alternatives' that retain some form of restraint of the sow (i.e. temporary crating).

The increased research and development activity in alternative systems is dominated by temporary crating designs (for reviews see Baxter et al., 2018; Glencorse et al., 2019; www.freefarrowing.org). Temporary crating systems (e.g. Table 4) facilitate loose-lactation but cannot support all the biological needs of the sow, especially her functionally important nest-building behaviour (Hansen et al., 2017). These systems are therefore a compromise; they are generally the most economically attractive choice out of all the alternatives (bar outdoor production – Guy et al., 2012). They are often built on the same spatial footprint as a conventional crate system, with fully or partly slatted flooring, and maintain most of the stockperson advantages that conventional systems impart (i.e. easy and safe intervention, easy dung removal). Where pledges to phase out farrowing crates have been made (e.g. Denmark and Austria - Hansen, 2018; Heidinger et al., 2018), there has been great investment in evaluating these compromise systems, including identifying when sows should be let out of their crates post-partum. The 'critical time window' for piglet survival has been suggested as four days post-partum (Heidinger et al., 2018). It can, of course, be argued that any reduction of restraint is an improvement in animal welfare, but temporary crating systems are far from optimal for sows or piglets. Many of the benefits of allowing for more satisfaction of nest-building motivation are not realised for the sow or her piglets, and ensuring the sow is released after a short period post-partum will be based on the stockperson's discretion. There can also be spikes in piglet mortality at this time if opening protocols are sub-optimal (King et al., 2019), or if pens are operated as zero-confinement when they have been designed for temporary crating. This is repeatedly seen in studies where the same design of temporary confinement system is compared when kept in an open position all of the time versus varying lengths of confinement (e.g. Hales et al., 2014; Condous et al., 2016; Lohmeier et al., 2020). As these systems rarely have all the design features that promote good maternal behaviour (as described above), it is almost inevitable that they will fail when the sow is completely free. It is understandable that temporary crating options may be perceived by farmers as a good introduction to loose-housing, with the crate offering an 'insurance' policy to return to conventional methods if there is no uptake either by staff or the supply chain. However, even temporary crating systems represent an investment - the most basic ones (i.e. those occupying the same footprint as conventional systems) cost ~1.6 times more than farrowing crates (Guy et al., 2012) and such investment might be costly if not future-proofed for consumer demands. This was evident in the poultry industry, where the introduction of the 'enriched (or furnished) cage' to replace battery cages did not fully appreciate public perception that 'a cage is still a cage' and therefore a push for alternatives which resulted in a surge in the free range market and development of barn egg and aviary alternatives.

Thus, it is still important to determine whether optimal welfare can be attained in a farrowing and lactation system where the needs of all stakeholders are taken into account (i.e. sow, piglet, stockperson and consumer).

3.4 Case study: PigSAFE - a zero confinement farrowing and lactation system

A project commissioned by the UK government aimed to develop a suitable alternative to the farrowing crate that provides for the maximal sow and piglet welfare that can be achieved under commercial conditions. This required the re-designing of a system that could reconcile a 'triangle of needs' relating to the farmer, the sow and her litter. However, as the needs of the litter and the farmer are often closely aligned, both prioritising improved piglet survival and growth, the main conflict to be resolved lies between the sow and the farmer. This requires methods to provide the appropriate level of environmental enrichment to meet the biological 'needs' of the farrowing sow that are also consistent with good piglet survival, and other management and business constraints. A literature review was undertaken to establish/confirm the biological principles (including ethological understanding) underlying sow and piglet needs in the perinatal period. From this, a list of biological specifications (or design criteria) for zero-confinement farrowing systems was derived (as discussed in summary above and in Table 1, and in detail in Baxter et al., 2011a). A second critical review was undertaken to gather information on all known alternative farrowing and lactation systems (including outdoor systems), to identify the extent to which each system addressed biological specifications and its practical success (Baxter et al., 2012; updated in Baxter et al., 2018). From this platform, and in collaboration with representatives from industry, non-governmental organisations, biologists, economists and engineers, a prototype system was developed - PigSAFE (Piglet and Sow Alternative Farrowing Environment). This was tested on two research farms to 'model' potential implementation scenarios (i.e. conversion of existing farrowing crate accommodation (Farm A) and a new-build (Farm B)) and to rapidly de-bug obvious problems.

3.4.1 The PigSAFE design

PigSAFE is a 'designed pen'. This terminology is used for a pen that has different areas to fulfil different functions (Baxter et al., 2012, 2018) and specific pen features designed to stimulate good animal behaviour (i.e. good maternal behaviours: correct farrowing location, careful lying behaviours, appropriate interactions; good hygiene: use of provided dunging sites). The PigSAFE designs on Farm A and B are shown in Figs 1a and 2a, respectively. They have similar features, but Farm A has a narrower dunging passage and is built above a slurry pit with plastic slotted and slatted flooring. Both pens have a nest area, with solid flooring to allow provision of nesting material, and sloping walls to control the stand-to-lie posture changes of the sow and lower the risk of piglets being trapped and killed. A heated creep area has easy access from the nest for the piglets and from the passageway for stockperson inspection. A separate slatted dunging area is bounded by walls with barred panels to adjacent pens, to discourage farrowing outside the nest and allow sow-to-sow visual and oral-nasal contact. A sow feeding crate (solid-sided) is included at one side of the pen, where the sow can be locked in to allow safe inspection or treatment of the piglets, but is not wide enough to allow the sow to be locked

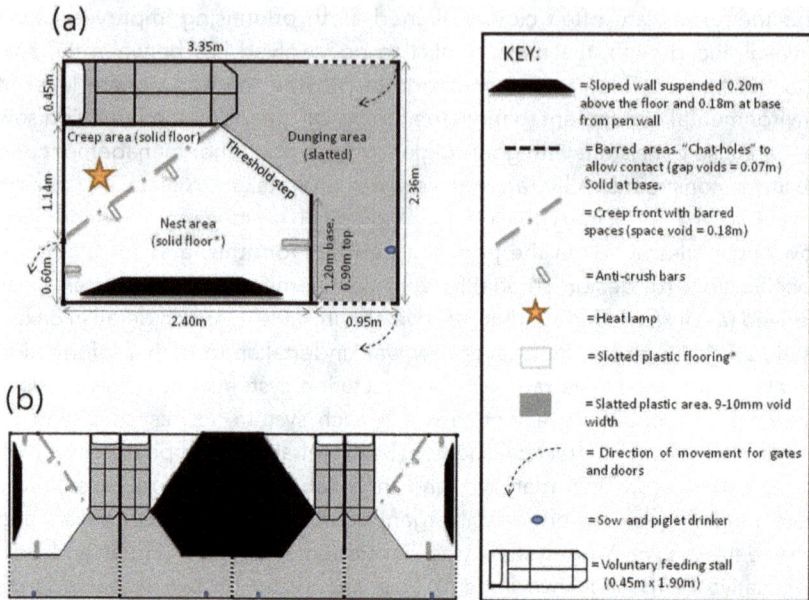

Figure 1 (a) Prototype PigSAFE pen (not for scale) and (b) Four pens were constructed in one room: in each room, two pens were provided with sound proofing material and a temporary roof fitted (height 1.2m) for the QUIET treatment.

in for farrowing. Farm B's new-build design incorporates additional dynamic features such as the ability to open up the nest area after a designated time (~7 days post-farrowing) to improve hygiene and simulate nest departure. The pen is partially slatted, with a scraper system under the metal slats in the dunging area.

The first iterations of both designs tested some gaps in the knowledge regarding the effects of a 'quiet' nest-site and quantification of substrate needed in the nest-site (Farm A – Edwards et al., 2012), as well as the effects of under-floor heating in the nest-site and quantification of nest size (Farm B – Baxter et al., 2015).

3.4.2 Initial trials

A concern when sows farrow loose is that they will not farrow in the desired location. In the PigSAFE design, that location was the nest-site where substrate was present and sloped walls provided protection and get-away sites for piglets, as well as a large heated creep. Initial trials were focussed on determining whether we could provide the correct stimuli to achieve the desired farrowing location and acceptable piglet survival (defined as comparable to farrowing crates and national herd averages).

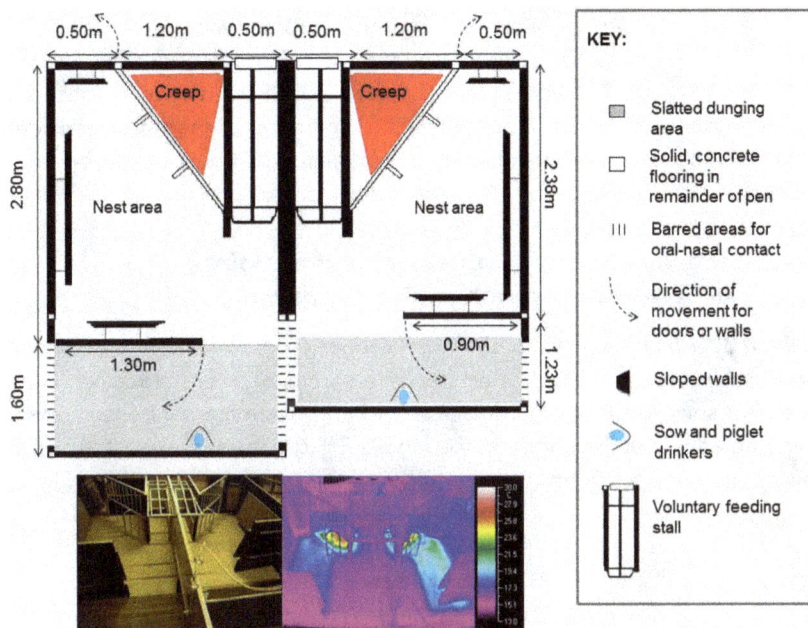

Figure 2 PigSAFE design showing LARGE (left) and SMALL (right) pen designs and thermal image showing dispersion of heat in the nest-site when under-floor heating was applied for treatments.

3.4.2.1 The influence of sound-proofing and substrate on farrowing location and performance

For the trials on Farm A, it was hypothesised that 1) 'ad libitum' substrate provided in the nest area would encourage nest-building and farrowing in the correct location, and 2) a quiet nest-site would encourage farrowing in the correct location and benefit piglet survival. The treatments were applied within the PigSAFE pen design (Fig. 1a) (overall pen size = 2.36m x 3.35m = 7.9m²). Sound-suppression material and a roof over the nest-site were used to create a 'quiet' nest (Fig. 1b), and long-stemmed straw was provided at two levels (MAX – 4 kg and MIN – 2 kg) before farrowing until two days post-farrowing. A 2 × 2 factorial design saw 99 Large White × Landrace sows randomly assigned to quiet (i.e. quiet or control) and substrate (MAX or MIN) treatments. Linear mixed models analysed data. All the sows chose to start farrowing in the nest and 99% of all piglets were farrowed in the nest. While there were numerical trends in the expected direction, neither the degree of nest enclosure/soundproofing nor the quantity of nesting substrate supplied had a statistically significant influence on piglet survival (Live-born mortality, Quiet = 14% vs. Control = 17%, sem = 1.52, P = 0.16; Min = 17% vs. Max = 14%, sem 1.52, P = 0.11).

The overall design successfully promoted farrowing in the desired location, irrespective of substrate amount or sound-suppression. This allowed the more cost-effective and practical pen design (i.e. no covered nest and sound proofing) and substrate amount (i.e. 2 kg long-stemmed straw provided in nest-site on entry and maintained at that level until farrowing) to be taken forward for the next iterations and 'commercial' trials.

3.4.2.2 The influence of temperature and nest size on farrowing location and performance

At Farm B, two hypotheses to optimise farrowing location and improve piglet survival were tested: 1) a heated nest-site would be more attractive to the farrowing sow, and 2) greater space would improve maternal behaviour and therefore benefit piglet performance. PigSAFE (Fig. 2) was adapted: the LARGE treatment measured 9.7m² in total, with a nest area of 4.0m², and the SMALL treatment (same design but smaller) measured 7.9m² in total with a nest area of 3.3m². The nest floor was heated to either 30°C (T30) or 20°C (T20) from 48h before, until 24h after farrowing. Room temperature was kept at 18°C.

A 2 × 2 factorial design saw 88 Large White × Landrace sows randomly assigned to space and temperature treatments. Generalised linear mixed models analysed data. The overall design successfully promoted farrowing in the desired location, irrespective of nest size and floor temperature (97% of all piglets born were born in the nest). There was no significant effect of floor temperature on performance. However, space influenced mortality, with significantly greater live-born mortality when sows were afforded a larger farrowing space (LARGE=18.1% vs. SMALL=10.9% P=0.028). Behavioural analysis suggested that the greater mortality was a result of a larger nest area in which the sow could lie down unsupported (full results of this study can be found in Baxter et al., 2015). These results suggest the larger nest size was less protective for the piglets and thus a smaller nest would be recommended. The pen features with nesting substrate provided enough stimuli, regardless of floor temperature, to attract sows into the nest.

Conclusion from the initial trials: the results indicated that the PigSAFE pen provided the correct stimuli (enclosed nest-site, substrate) to attract the sow to farrow in the desired location and that piglet survival levels were acceptable at both sites within the constraints of the experiment. The final prototypes would adopt the best designs at both sites (which were also the most cost-effective) to run 'commercially' using farm rather than research staff.

3.4.3 PigSAFE 'commercial' trials and performance

For this phase, farm staff at each site were responsible for operating both PigSAFE and farrowing crates in parallel, given minimal experimental constraints.

The optimised designs were used at each site and only basic instructions were given, including the best practice for substrate provision based on initial trials. Data presented are on pig performance.

3.4.3.1 Piglet survival

Combined data from the two sites (304 sows, n=164 in Crates n=140 in PigSAFE, approximately 10% gilts) are shown in Table 2 against the national average and top third data at the time of the trial.

When the data at each farm were looked at separately, there were interesting trends suggesting farm differences. At Farm A, the PigSAFE system performed very well, with no significant difference compared to crates (Fig. 3) and both systems performing better than the industry top third. At Farm B, although there were no significant differences in pig performance, mortality in the PigSAFE pen was closer to the industry indoor average.

At Farm B, pig performance in PigSAFE pens showed an improvement with every batch farrowed (Fig. 4), highlighting that stock-person training is an important aspect when adopting a new system. In contrast with Farm A, where the manager had previous experience with loose-farrowing outdoor sows, the manager at Farm B had previously worked with sows in crates for over 25 years and thus was unfamiliar with loose farrowing and lactating sows. The learning curve is indicative of potential scenarios when implementing a new system on a commercial farm where staff have previously had little or no experience of animals farrowing and lactating while loose.

Table 2 Performance of PigSAFE pens and farrowing crates for the period up to August 2011, with national average and top third farrowing crate data for comparison (AHDB Pork, 2010[1])

	Total born (BA+BD)	Total weaned	%Total mortality	%Live-born mortality	%Stillborn mortality*
PigSAFE	13.0	10.7[a]	13.3[b]	9.6[c]	4.1[b]
Farrowing crates	12.8	10.8[a]	12.0[b]	9.0[c]	3.4[b]
National indoor average	12.3	10.1	17.6	12.3	6.1
National indoor top third	13.4	11.2	16.5	10.3	5.6

[a]Adjusted for net fostering; [b]This figure includes both stillborn piglets and those dying before weaning and is adjusted for litter size; [c]Adjusted for litter size post fostering. BA = Born alive, BD = Born dead/Stillborn. *Stillborn% = percentage of total litter size that were born dead.
[1] https://porktools.ahdb.org.uk/prices-stats/costings-herd-performance/indoor-breeding-herd/.

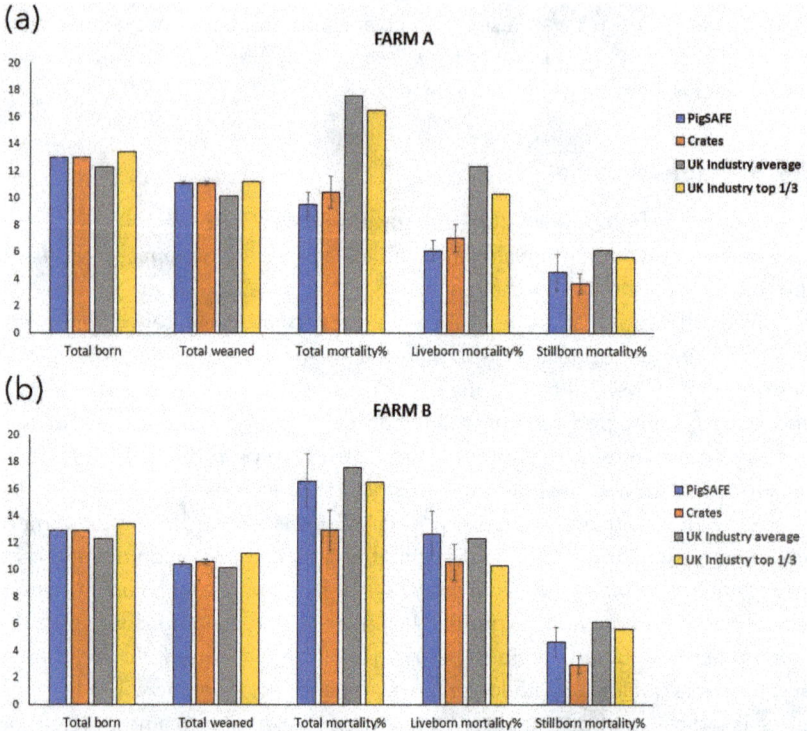

Figure 3 Performance at Farm A (a) and Farm B (b). Descriptive comparison against industry average and top third at the time (AHDB Pork, 2010, https://porktools.ahdb.or g.uk/prices-stats/costings-herd-performance/indoor-breeding-herd/). Mortality data are adjusted for net fostering (similar in both systems), parity and litter size.

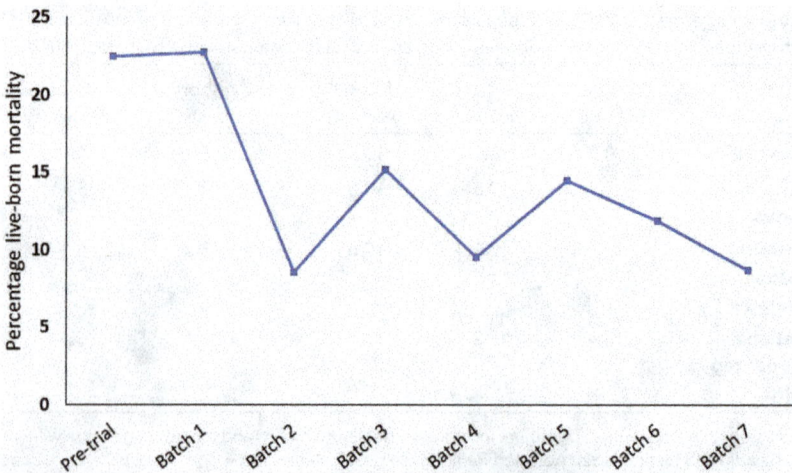

Figure 4 Effect of staff experience on live-born mortality in the PigSAFE system at Farm B.

3.4.3.2 Sow and piglet weight and condition

Body weight (kg) and back-fat depth were measured on sows pre-farrowing and post-weaning, and piglets were weighed at weaning. Table 3 shows that although there were no significant differences between housing systems at Farm A, at Farm B piglets weaned from PigSAFE pens were significantly heavier than those from crates (average individual weight 8.8 kg vs. 8.5 kg).

At Farm B there was a tendency for sows housed in PigSAFE pens to eat more than those housed in crates (7.27 kg vs. 6.44 kg per sow per day $F_{1,6}$=3.45 P=0.088). At Farm A there were no significant differences in feed intake between farrowing systems (P=0.621). It is possible that the difference in feed intake is only seen at Farm B as a result of potentially beneficial effects of the building in which the PigSAFE pens were built. The 'new build' scenario implemented at Farm B involved a large building shell with high roof space, and thus a more airy environment for the sows. It is possible that the higher piglet weaning weights seen in PigSAFE at Farm B relate to this sow feed intake, but it could also be that the more open pens (at 7 days post-farrowing the dividing wall opened up) improved udder access and milk let-down. This has been seen in other work demonstrating the benefits of space on suckling success as a result of easier udder access (Pedersen et al., 2011). These performance advantages are important, as the capital investment required to install a PigSAFE pen has been shown to be more than a conventional crate (Guy et al., 2012).

Conclusions from commercial-type trials: Results from the commercial scale-up phase showed no significant differences in performance between PigSAFE pens and conventional crates within farms. The most promising results from Farm A showed that the PigSAFE system can perform better than the industry top third producers. A key aspect of any new system is the ability and willingness of stockpersons to adapt to a new system and this effect was clearly

Table 3 Influence of farrowing accommodation on sow body condition and piglet weaning weight (adjusted for litter size, weaned litter size and weaning age). Sow weight loss includes the birth of the litter. For statistical purposes, systems within sites were compared.

	Sow condition		Litter condition
	Weight loss (kg)	Back-fat loss (mm)	Average litter weaned weight (kg)
Farm A PigSAFE	38.81	4.18	78.05
Farm A Farrowing crates	38.54	4.28	76.14
Farm B PigSAFE	27.68	4.45	91.33[a]
Farm B Farrowing crates	29.71	4.18	87.77[b]

Subscripts with different letters indicate figures are significantly different at P < 0.05.

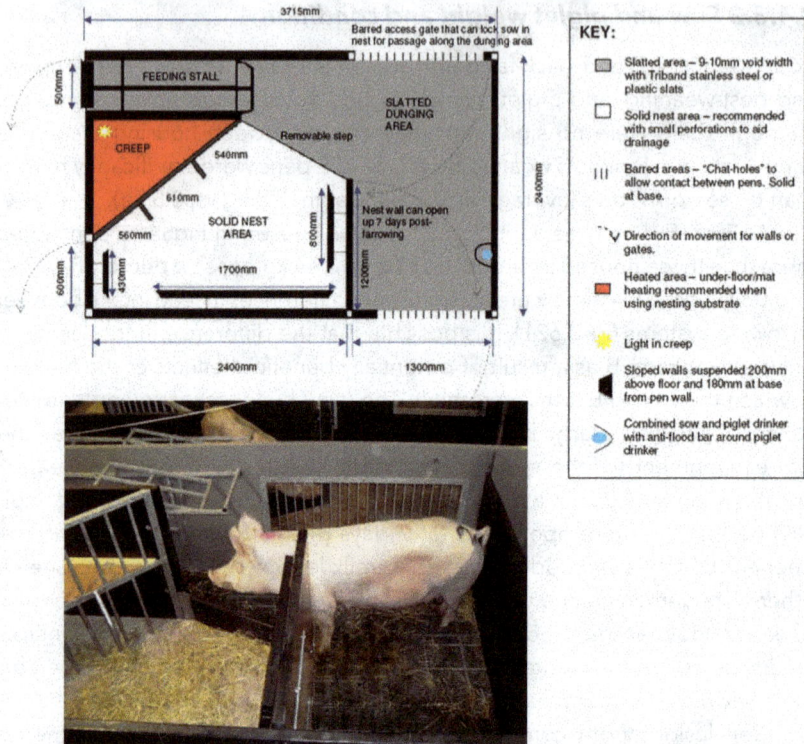

Figure 5 Recommended design and dimensions for building a PigSAFE pen.

seen at Farm B. There is also some indication of benefits in weaning weights and sow feed intake when using PigSAFE pens compared with conventional crates. The recommended best design as a result of the research is shown in Fig. 5, and full details of all components can be found at 'www.freefarrowing .org'.

Throughout the project it was emphasised that the detail of the design is very important in its performance and therefore any deviation from the recommended details is done at the farmer's own risk.

3.5 Commercial developments in alternative farrowing and lactation systems

Since the publication of the initial review manuscripts (Baxter et al., 2011a, 2012), there has been a great deal of research and industry activity in the development of alternative systems (Table 4). These have included large national studies in Austria (Pro-SAU) and Denmark (SEGES showroom) and plans for reform to phase out farrowing crates in these countries. Thus Baxter

Table 4 Examples of different farrowing and lactation systems (Specific information on alternative systems can be found at www.freefarrowing.org)

System		Descriptions	
Crates	Conventional	Crate	Tubular metal bars run horizontally along the length of the crate, with additional bars positioned above the sow to prevent escape via climbing or jumping. Crate prevents turning around. Crates are fixed width or can be adjusted at the back for larger sows. Space available to sow: 0.90–1.23 m². Total space available to piglets: ~3.5–4.0 m². A water and food trough is situated at the front with piglet water drinker within pen. Flooring: fully or part-slatted with pens usually having a solid floor or mat for the piglet resting area, which includes a heat source.	Fixed-width farrowing crate with front covered and heated creep, solid floors with cast-iron slatted area at rear for waste removal. Photo courtesy of E. M. Baxter. Adjustable farrowing crate with capability to widen but prevent turning. Corner, heated creep, partially slatted floors (concrete/cast-iron). Photo courtesy of V. A. Moustsen.

(Continued)

Table 4 (*Continued*)

System		Descriptions	
Temporary crating	Hinged/ swing-side for example, 360° MPP, Combi-flex, Combi-flex hexel, Winged	Farrowing crate structure within pen as in conventional systems. Spatial footprint and flooring same as conventional system. Sow is crated on entry to farrowing house until approximately 5–7 days post-farrowing. Crate can be opened up or swung-open for lactation allowing the sow to turn around. An additional heated and solid area (typically a heat-mat) inaccessible to the sow is provided for the piglets.	360° MPP temporary crate in 'open' (left) and 'closed' (right) position. Crate typically opened after 4–7 days. Heat mat for piglets on fully slatted plastic floor. Photo courtesy of E. M. Baxter. 'Winged' temporary crating option. Crate opens up at the back to allow the sow to reverse out and turn in the dunging area. Photo courtesy of V. A. Moustsen.

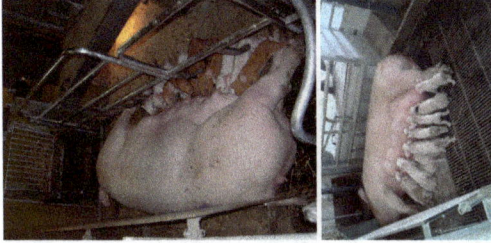

SWAP in closed and open positions. Photo courtesy of V. A. Moustsen.

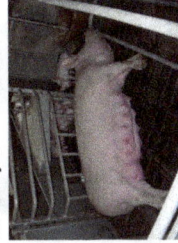

Simple pen on the same spatial footprint as a conventional farrowing crate. Fully slatted flooring. Photo courtesy of V. A. Moustsen.

(Continued)

Pens	SWAP	The SWAP system (Sow Welfare And Piglet Protection) is designed to allow nest-building and then confinement post-farrowing until approximately four days post-farrowing (Moustsen et al., 2013). Sow should be able to turn around for nesting and lactation but crated for farrowing and immediately post-partum. Larger footprint than the crate – 6 m^2 built on part-slatted flooring. Separate large corner heated and covered area for piglets.
	Simple	For example, simple pen, mushroom pen, sloped floor	Same spatial footprint as conventional crate. No crate. Sow able to turn around at all times. Fully or part-slatted flooring. Possibility of modified floor design. Separate heated area for piglets.

Table 4 Examples of different farrowing and lactation systems (Specific information on alternative systems can be found at www.freefarrowing.org) (Continued)

System		Descriptions	
Designed pen	For example, Comfort Sow, Danish FF, FATs, PigSAFE, WelCon, Werribee	Zero-confinement. Sow can turn around. Defined areas provided in the pen for feeding, dunging and lying/nesting. Size varies (5–8.5 m²). Additional pen 'furniture' such as rails or sloped walls to assist sow posture changes and protect piglets. Solid or part-slatted floor with generous lying area to provide substrate for nest-building. A separate, heated creep area for the piglets, inaccessible to the sow is provided. Some systems have the potential for separating the sow from piglets to allow farmers to perform husbandry procedures safely.	Danish Free Farrower Zero confinement pen with part-slatted flooring and a corner creep. Photo courtesy of V. A. Moustsen. PigSAFE zero confinement pen with part slatted flooring, corner creep and separate feeding, nesting and dunging areas. Photo courtesy of E. M. Baxter.

Group	Zero-confinement group	Separate or free-access nests then group	Systems allow sows and litters to mix before weaning, typically 10–21 days into lactation. Based on multi-suckling accommodation (e.g. Swedish systems – Ljungström, Thorstensson). Variable but large amounts of space for both sows and piglets. For farrowing, sows are initially individually housed in single pens, but are integrated with their litter into groups in larger multi-suckling pens between 10–21 days post-farrowing. Alternatively, sows are already grouped prior to farrowing and have free access to individual nest boxes for farrowing.	

Typical Swedish multi-suckling system with deep-straw bedding. Photo courtesy of J. H. Pedersen.

Group housing system for lactating sows at Swine Innovation Center Sterksel in the Netherlands.
Photo courtesy of C. van der Peet-Schwering, Wageningen University and Research Centre.

(Continued)

Table 4 Examples of different farrowing and lactation systems (Specific information on alternative systems can be found at www.freefarrowing.org) (Continued)

System			Descriptions	
Kennels	Temporary crating then group	Crated then grouped	Sows are conventionally crated and then grouped 10–14 days post-partum to allow litter integration.	
	Kennels	Kennel with run, for example, 'Solari'	An outside space is intended for dunging and feeding by the individually housed sow, with an indoor kennel for farrowing. Floors are solid to facilitate provision of deep substrate. A heated, creep space may be provided within the kennel for the piglets.	Outdoor kennel and run. Photo courtesy of E. M. Baxter.
Outdoor	Outdoor	Outdoor	These are systems with low capital investment and running costs, where sows and their piglets are housed individually, outdoors in farrowing arks or huts, with access to individual or group paddocks. There are different ark and hut designs available and described in detail elsewhere (e.g. Honeyman et al., 1998; Baxter et al., 2009).	Outdoor farrowing huts and paddocks. Photos courtesy of S.-L. A. Schild (top) and E. M. Baxter (bottom).

et al. (2018) updated their earlier scientific review to include new information in the state-of-the-art. In addition, Baxter and Edwards developed a website to provide a more farmer-friendly interface for those interested in building alternative systems (www.freefarrowing.org).

Many of the alternative systems available commercially are temporary crating systems, as farmers seek to retain greater control over the sow's movements, particularly around the first few days post-partum when interventions to promote piglet survival and piglet husbandry procedures are performed. Farmers have also said that they want an 'insurance policy' of being able to resort to the conventional crated system if they do not succeed in adopting free farrowing or loose lactation, either because of stockperson reluctance, performance problems or a failure in the market to recognise the additional financial requirements of higher welfare systems. Although there has been a lot of work on pen design, less attention has been given to optimising management and genetic selection of the best animals for loose indoor systems. It is likely that all three 'Ps' (Pens-People-Pigs) will need to be optimised to achieve consistent performance and reduce the barriers to uptake.

4 Conclusions

Determining the basic needs of all actors (animals, farmers) that are involved in a system is a key starting point for designing higher welfare, sustainable alternatives to conventional farrowing crate systems that are known to impose welfare challenges for both sows and piglets.

Allowing the animals to be more active in the control of their environment is a positive aspect of animal welfare that appears to be an important part of their behavioural needs. When given the chance to be an active participant in her own environment, the sow will fulfil her own needs but also contribute to the goals of others. Maximising piglet survival increases her biological fitness and also satisfies the main needs of the piglets and farmer. The current challenge is what happens when the biological limits of the sow have been exceeded and she can no longer fully control the outcomes of all her offspring without the contribution of another actor. Designing systems that can allow stockperson interventions without imposing any further constraints on sow welfare is an important goal for research and development.

5 Future trends in research

This chapter has concentrated on the managerial and environmental interventions required to optimise animal welfare around farrowing. Research on the development of alternative farrowing and lactation systems is reasonably mature; however, greater innovation is required to optimise system components. For example, designing suitable flooring and waste management systems that allow the provision of environmental enrichment without compromising hygiene is still a challenge.

Optimising the biological components of farrowing and lactation systems will involve breeding for improved maternal behaviour (Grandison, 2005; Gäde et al., 2008; Baxter et al., 2011b) and investigating strategies for breeding a more robust piglet. Both have potential to improve piglet survival and therefore the potential to promote greater confidence with loose housing. However, breeding strategies seeking short-term gains in production, such as selecting for increased numbers of piglets born beyond the biological limits of the sow to rear them, threaten welfare. They also challenge system design to incorporate features that help cater for the increasingly specialised needs of piglets born into large litters. Finally, a social science element meriting further research is understanding the barriers to uptake of alternative systems, including stockperson behaviour and willingness to change. Greater attention to all these aspects should improve sow and piglet welfare while addressing farmer concerns.

6 Where to look for further information

In addition to the peer-reviewed literature provided in the reference section, web-based information can be found detailing practical ways to optimise the farrowing and lactation environment.

- www.freefarrowing.org.
- http://practicalpig.ahdb.org.uk/.

7 References

Ahlström, S., Jarvis, S. and Lawrence, A. B. 2002. Savaging gilts are more restless and more responsive to piglets during the expulsive phase of parturition. *Applied Animal Behaviour Science* 76(1), 83–91.

Alexopoulos, J. G., Lines, D. S., Hallett, S. and Plush, K. J. 2018. A review of success factors for piglet fostering in lactation. *Animals* 8(3), 38.

Algers, B. and Uvnäs-Moberg, K. 2007. Maternal behavior in pigs. *Hormones and Behavior* 52(1), 78–85.

Amdi, C., Hansen, C. F., Krogh, U., Oksbjerg, N. and Theil, P. K. 2013a. Less brain sparing occurs in severe intrauterine growth-restricted piglets born to sows fed palm fatty acid distillate. In: *Manipulating Pig Production XIV*. APSA Biennial Conference, Melbourne, Victoria, Australia, 24th to 27th November, p. 125.

Amdi, C., Krogh, U., Flummer, C., Oksbjerg, N., Hansen, C. F. and Theil, P. K. 2013b. Intrauterine growth restricted piglets defined by their head shape ingest insufficient amounts of colostrum. *Journal of Animal Science* 91(12), 5605–5613.

Andersen, H. M. L. and Pedersen, L. J. 2011. The effect of feed trough position on choice of defecation area in farrowing pens by loose sows. *Applied Animal Behaviour Science* 131(1–2), 48–52.

Andersen, I. L., Vasdal, G. and Pedersen, L. J. 2014. Nest building and posture changes and activity budget of gilts housed in pens and crates. *Applied Animal Behaviour Science* 159, 29-33.

Arey, D. S. 1992. Straw and food as reinforcers for prepartal sows. *Applied Animal Behaviour Science* 33(2-3), 217-226.

Balzani, A., Cordell, H. J., Sutcliffe, E. and Edwards, S. A. 2016. Heritability of udder morphology and colostrum quality traits in swine. *Journal of Animal Science* 94(9), 3636-3644.

Bandrick, M., Pieters, M., Pijoan, C., Baidoo, S. K. and Molitor, T. W. 2011. Effect of cross-fostering on transfer of maternal immunity to Mycoplasma hyopneumoniae to piglets. *Veterinary Record* 168(4), 100.

Barnett, J. L., Hemsworth, P. H., Cronin, G. M., Jongman, E. C. and Hutson, G. D. 2001. A review of the welfare issues for sows and piglets in relation to housing. *Australian Journal of Agricultural Research* 52(1), 1-28.

Baxter, E. M. and Edwards, S. A. 2018. Piglet mortality and morbidity: inevitable or unacceptable? In: Špinka, M. (Ed.), *Advances in Pig Welfare*, pp. 73-100. Woodhead Publishing, Cambridge, UK.

Baxter, E. M., Jarvis, S., D'eath, R. B., Ross, D. W., Robson, S. K., Farish, M., Nevison, I. M., Lawrence, A. B. and Edwards, S. A. 2008. Investigating the behavioural and physiological indicators of neonatal survival in pigs. *Theriogenology* 69(6), 773-783.

Baxter, E. M., Jarvis, S., Sherwood, L., Robson, S. K., Ormandy, E., Farish, M., Smurthwaite, K. M., Roehe, R., Lawrence, A. B. and Edwards, S. A. 2009. Indicators of piglet survival in an outdoor farrowing system. *Livestock Science* 124(1-3), 266-276.

Baxter, E. M., Lawrence, A. B. and Edwards, S. A. 2011a. Alternative farrowing systems: design criteria for farrowing systems based on the biological needs of sows and piglets. *Animal: An International Journal of Animal Bioscience* 5(4), 580-600.

Baxter, E. M., Jarvis, S., Sherwood, L., Farish, M., Roehe, R., Lawrence, A. B. and Edwards, S. A. 2011b. Genetic and environmental effects on piglet survival and maternal behaviour of the farrowing sow. *Applied Animal Behaviour Science* 130(1-2), 28-41.

Baxter, E. M., Lawrence, A. B. and Edwards, S. A. 2012. Alternative farrowing accommodation: welfare and economic aspects of existing farrowing and lactation systems for pigs. *Animal* 6(1), 96-117.

Baxter, E. M., Rutherford, K. M. D., d'Eath, R. B., Arnott, G., Turner, S. P., Sandøe, P., Moustsen, V. A., Thorup, F., Edwards, S. A. and Lawrence, A. B. 2013. The welfare implications of large litter size in the domestic pig II: management factors. *Animal Welfare* 22(2), 219-238.

Baxter, E. M., Adeleye, O. O., Jack, M. C., Farish, M., Ison, S. H. and Edwards, S. A. 2015. Achieving optimum performance in a loose-housed farrowing system for sows: the effects of space and temperature. *Applied Animal Behaviour Science* 169, 9-16.

Baxter, E. M., Andersen, I. L. and Edwards, S. A. 2018. Sow welfare in the farrowing crate and alternatives. In: Špinka, Marek (Ed.), *Advances in Pig Welfare*, pp. 27-72. Woodhead Publishing, Cambridge, UK.

Baxter, M. R. and Schwaller, C. E. 1983. Space requirements for sows in confinement. In: Baxter, S. H., Baxter, M. R. and MacCormack, J. (Eds), *Farm Animal Housing and Welfare*, pp. 181-195. Martinus Nijhoff Publishers, The Hague, The Netherlands.

Baxter, M. R. 1991. The Freedom farrowing system. *Farm Building Progress* 104, 9-15.

Björkman, S., Oliviero, C., Rajala-Schultz, P. J., Soede, N. M. and Peltoniemi, O. A. T. 2017. The effect of litter size, parity and farrowing duration on placenta expulsion and retention in sows. *Theriogenology* 92, 36–44.

Black, J. L., Mullan, B. P., Lorschy, M. L. and Giles, L. R. 1993. Lactation in the sow during heat stress. *Livestock Production Science* 35(1-2), 153–170.

Bøe, K. E., Kvaal, I., Hall, E. J. S. and Cronin, G. M. 2016. Individual differences in dunging patterns in loose-housed lactating sows. *Acta Agriculturae Scandinavica, Section A– Animal Science* 66(4), 221–230.

Bohnenkamp, A. L., Traulsen, I., Meyer, C., Müller, K. and Krieter, J. 2013. Comparison of growth performance and agonistic interaction in weaned piglets of different weight classes from farrowing systems with group or single housing. *Animal: An International Journal of Animal Bioscience* 7(2), 309–315.

Bolhuis, J. E., Raats-van den Boogaard, A. M. E., Hoofs, A. I. J. and Soede, N. M. 2018. Effects of loose housing and the provision of alternative nesting material on peri-partum sow behaviour and piglet survival. *Applied Animal Behaviour Science* 202, 28–33.

Bonde, M. 2008. Prevalence of decubital shoulder lesions in Danish sow herds. Internal Report 12, University of Aarhus, Faculty of Agricultural Sciences, p. 8.

Bontempo, V. and Jiang, X. R. 2015. Feeding various fat levels in sows: effects on immune status and performance of sows and piglets. In: Farmer, C. (Ed.), *The Gestating and Lactating Sow*, pp. 357–375. Wageningen Academic Publishers, Netherlands.

Boogaard, B. K., Boekhorst, L. J. S., Oosting, S. J. and Sørensen, J. T. 2011. Socio-cultural sustainability of pig production: citizen perceptions in the Netherlands and Denmark. *Livestock Science* 140(1-3), 189–200.

Bos, B. and Koerkamp, P. G. 2009. Synthesising needs in system innovation through structured dresign: a methodical outline of the role of needs in reflexive interactive design (RIO). Sustainable agriculture and food chains in peri-urban areas. In: Poppe, K. J., Termeer, K. and Slingerland, M. (Eds), *Transitions Towards Sustainable Agriculture and Food Chains in Peri-Urban Areas*, Chapter 12p. 219–237. Wageningen Academic Publishers.

Boyle, L. A., Leonard, F. C., Lynch, P. B. and Brophy, P. 2002. Effect of gestation housing on behaviour and skin lesions of sows in farrowing crates. *Applied Animal Behaviour Science* 76(2), 119–134.

Boyle, L. A., Regan, D., Leonard, F. C., Lynch, P. B. and Brophy, P. 2000. The effect of mats on the welfare of sows and piglets in the farrowing house. *Animal Welfare* 9(1), 39–48.

Brown, S. M., Klaffenböck, M., Nevison, I. M. and Lawrence, A. B. 2015. Evidence for litter differences in play behaviour in pre-weaned pigs. *Applied Animal Behaviour Science* 172, 17–25.

Buller, H., Blokhuis, H., Jensen, P. and Keeling, L. 2018. Towards farm animal welfare and sustainability. *Animals* 8(6), 81.

Cabrera, R. A., Boyd, R. D., Jungst, S. B., Wilson, E. R., Johnston, M. E., Vignes, J. L. and Odle, J. 2010. Impact of lactation length and piglet weaning weight on long-term growth and viability of progeny. *Journal of Animal Science* 88(7), 2265–2276.

Calderón Díaz, J. A., Fahey, A. G., KilBride, A. L., Green, L. E. and Boyle, L. A. 2013. Longitudinal study of the effect of rubber slat mats on locomotory ability, body, limb and claw lesions, and dirtiness of group housed sows. *Journal of Animal Science* 91(8), 3940–3954.

Calderón Díaz, J. A., Diana, A., Boyle, L. A., Leonard, F. C., McElroy, M., McGettrick, S., Moriarty, J. and García Manzanilla, E. 2017. Delaying pigs from the normal production flow is associated with health problems and poorer performance. *Porcine Health Management* 3(1), 13.

Chen, J. and Lobo, A. 2012. Organic food products in China: determinants of consumers' purchase intentions. *The International Review of Retail, Distribution and Consumer Research* 22(3), 293-314.

Condous, P. C., Plush, K. J., Tilbrook, A. J. and Van Wettere, W. H. E. J. 2016. Reducing sow confinement during farrowing and in early lactation increases piglet mortality. *Journal of Animal Science* 94(7), 3022-3029.

Council Directive 2008/120/EC of 18 December 2008 laying down minimum standards for the protection of pigs.

Courboulay, V., le Roux, A., Collin, F., Dutertre, C. and Rousseau, P. 2000. Incidence du type de sol en maternite sue le confort de la truie et des porcelets. *Journees Recherche Porcine en France* 32, 115-122.

Cronin, G. M., Barnett, J. L., Hodge, F. M., Smith, J. A. and McCallum, T. H. 1991. The welfare of pigs in two farrowing/lactation environments: cortisol responses of sows. *Applied Animal Behaviour Science* 32(2-3), 117-127.

Cronin, G. M., Smith, J. A., Hodge, F. M. and Hemsworth, P. H. 1994. The behaviour of primiparous sows around farrowing in response to restraint and straw bedding. *Applied Animal Behaviour Science* 39(3-4), 269-280.

Damm, B. I., Vestergaard, K. S., Schroder-Petersen, D. L. and Ladewig, J. 2000. The effect of branches on prepartum nest building in gilts with access to straw. *Applied Animal Behaviour Science* 69(2), 113-124.

Damm, B. I., Moustsen, V., Jørgensen, E., Pedersen, L. J., Heiskanen, T. and Forkman, B. 2006. Sow preferences for walls to lean against when lying down. *Applied Animal Behaviour Science* 99(1-2), 53-63.

D'Eath, R. B., Tolkamp, B. J., Kyriazakis, I. and Lawrence, A. B. 2009. 'Freedom from hunger' and preventing obesity: the animal welfare implications of reducing food quantity or quality. *Animal Behaviour* 77(2), 275-288.

de Barcellos, M. D., Grunert, K. G., Zhou, Y., Verbeke, W., Perez-Cueto, F. J. A. and Krystallis, A. 2013. Consumer attitudes to different pig production systems: a study from mainland China. *Agriculture and Human Values* 30(3), 443-455.

Decaluwé, R., Maes, D., Wuyts, B., Cools, A., Piepers, S. and Janssens, G. P. J. 2014. Piglets Colostrum intake associates with daily weight gain and survival until weaning. *Livestock Science* 162, 185-192.

Declerck, I., Dewulf, J., Piepers, S., Decaluwé, R. and Maes, D. 2015. Sow and litter factors influencing colostrum yield and nutritional composition. *Journal of Animal Science* 93(3), 1309-1317.

De Jonge, F. H., Bokkers, E. A. M., Schouten, W. G. P. and Helmond, F. A. 1996. Rearing piglets in a poor environment: developmental aspects of social stress in pigs. *Physiology and Behavior* 60(2), 389-396.

Devillers, N., Le Dividich, J. and Prunier, A. 2011. Influence of colostrum intake on piglet survival and immunity. *Animal: An International Journal of Animal Bioscience* 5(10), 1605-1612.

De Vos, M., Huygelen, V., Willemen, S., Fransen, E., Casteleyn, C., Van Cruchten, S., Michiels, J. and Van Ginneken, C. 2014. Artificial rearing of piglets: effects on small intestinal morphology and digestion capacity. *Livestock Science* 159, 165-173.

Donovan, T. S. and Dritz, S. S. 2000. Effect of split nursing on variation in pig growth from birth to weaning. *Journal of the American Veterinary Medical Association* 217(1), 79–81.

Douglas, S. L., Edwards, S. A., Sutcliffe, E., Knap, P. W. and Kyriazakis, I. 2013. Identification of risk factors associated with poor lifetime growth performance in pigs. *Journal of Animal Science* 91(9), 4123–4132.

Douglas, S. L., Edwards, S. A. and Kyriazakis, I. 2014. Management strategies to improve the performance of low birth weight pigs to weaning and their long term consequences. *Journal of Animal Science* 92(5), 2280–2288.

Drake, A., Fraser, D. and Weary, D. M. 2008. Parent-offspring resource allocation in domestic pigs. *Behavioral Ecology and Sociobiology* 62(3), 309–319.

D'Silva, J. and Turner, J. (Eds) 2012. *Animals, Ethics and Trade: The Challenge of Animal Sentience*. Earthscan, London, UK.

Duncan, I. J. 2005. Science-based assessment of animal welfare: farm animals. *Revue Scientifique et Technique* 24(2), 483–492.

Edwards, S. A. 2002. Perinatal mortality in the pig: environmental or physiological solutions? *Livestock Production Science* 78(1), 3–12.

Edwards, S. A., Brett, M., Seddon, Y. M., Ross, D. and Baxter, E. M. 2012. Evaluation of nest design and nesting substrate options for the PigSAFE free farrowing pen. *Proceedings of the British Society of Animal Science*, 7.

Edwards, S. A., Matheson, S. M. and Baxter, E. M. 2019a. Genetic influences on intra-uterine growth retardation of piglet and management interventions for low birth weight piglets. In: González, E. R. (Ed.), *Nutrition of Hyperprolific Sows*, pp. 141–166. Novus International, Inc., Spain.

Edwards, L. E., Plush, K. J., Ralph, C. R., Morrison, R. S., Acharya, R. Y. and Doyle, R. E. 2019b. Enrichment with Lucerne hay improves sow maternal behaviour and improves piglet survival. *Animals* 9(8), 558.

Engelsmann, M. N., Hansen, C. F., Nielsen, M. N., Kristensen, A. R. and Amdi, C. 2019. Glucose injections at birth, warmth and placing at a nurse sow improve the growth of IUGR piglets. *Animals* 9(8), 519.

Eurobarometer 2016. *Special Eurobarometer 442./Wave EB84.4 – TNS Opinion & Social. Attitudes of Europeans towards Animal Welfare*, Available at: http://ec.europa.eu/COMMFrontOffice/PublicOpinion/index.cfm/ResultDoc/download/Document.

Feldpausch, J. A., Jourquin, J., Bergstrom, J. R., Bargen, J. L., Bokenkroger, C. D., Davis, D. L., Gonzalez, J. M., Nelssen, J. L., Puls, C. L., Trout, W. E. and Ritter, M. J. 2019. Birth weight threshold for identifying piglets at risk for preweaning mortality. *Translational Animal Science* 3(2), 633–640.

Feyera, T., Pedersen, T. F., Krogh, U., Foldager, L. and Theil, P. K. 2018. Impact of sow energy status during farrowing on farrowing kinetics, frequency of stillborn piglets, and farrowing assistance. *Journal of Animal Science* 96(6), 2320–2331.

Fox, C., Merali, Z. and Harrison, C. 2006. Therapeutic and protective effect of environmental enrichment against psychogenic and neurogenic stress. *Behavioural Brain Research* 175(1), 1–8.

Gäde, S., Bennewitz, J., Kirchner, K., Looft, H., Knap, P. W., Thaller, G. and Kalm, E. 2008. Genetic parameters for maternal behaviour traits in sows. *Livestock Science* 114(1), 31–41.

Gaskin, H. R. and Kelly, K. W. 1995. Immunology and neonatal Mortality. In: Varley, M. A. (Ed.), *The Neonatal Pig: Development and Survival*, pp. 39–56. CAB International, UK.

Glencorse, D., Plush, K., Hazel, S., D'Souza, D. and Hebart, M. 2019. Impact of non-confinement accommodation on farrowing performance: A systematic review and meta-analysis of farrowing crates Versus pens. *Animals* 9(11), 957.

Gonyou, H. W., Brumm, M. C., Bush, E., Deen, J., Edwards, S. A., Fangman, T., McGlone, J. J., Meunier-Salaun, M., Morrison, R. B., Spoolder, H., Sundberg, P. L. and Johnson, A. K. 2006. Application of broken-line analysis to assess floor space requirements of nursery and grower-finisher pigs expressed on an allometric basis. *Journal of Animal Science* 84(1), 229-235.

Grandinson, K. 2005. Genetic background of maternal behaviour and its relation to offspring survival. *Livestock Production Science* 93(1), 43-50.

Greenwood, E. C., van Dissel, J., Rayner, J., Hughes, P. E. and van Wettere, W. H. E. J. 2019. Mixing sows into alternative lactation housing affects sow aggression at mixing, future reproduction and piglet injury, with marked differences between multisuckle and sow separation systems. *Animals* 9(9), 658.

Grunert, K. G., Sonntag, W. I., Glanz-Chanos, V. and Forum, S. 2018. Consumer interest in environmental impact, safety, health and animal welfare aspects of modern pig production: results of a cross-national choice experiment. *Meat Science* 137, 123-129.

Guy, J. H., Cain, P. J., Seddon, Y. M., Baxter, E. M. and Edwards, S. A. 2012. Economic evaluation of high welfare indoor farrowing systems for pigs. *Animal Welfare* 21(1) (Suppl.1), 19-24.

Hales, J., Moustsen, V. A., Nielsen, M. B. F. and Hansen, C. F. 2014. Higher preweaning mortality in free farrowing pens compared with farrowing crates in three commercial pig farms. *Animal* 8(1), 113-120.

Hales, J., Moustsen, V. A., Devreese, A. M., Nielsen, M. B. F. and Hansen, C. F. 2015. Comparable farrowing progress in confined and loose housed hyper-prolific sows. *Livestock Science* 171, 64-72.

Hamilton, W. D. 1964a. The genetical evolution of social behaviour. I. *Journal of Theoretical Biology* 7(1), 1-16.

Hamilton, W. D. 1964b. The genetical evolution of social behaviour. II. *Journal of Theoretical Biology* 7(1), 17-52.

Hansen, C. F., Hales, J., Weber, P. M., Edwards, S. A. and Moustsen, V. A. 2017. Confinement of sows 24 h before expected farrowing affects the performance of nest building behaviours but not progress of parturition. *Applied Animal Behaviour Science* 188, 1-8.

Hansen, L. U. 2018. Test of 10 different farrowing pens for loose-housed sows. Report no. 1803. SEGEs, Denmark.

Hartsock, T. G. and Barczewski, R. A. 1997. Prepartum behaviour in swine: effects of pen size. *Journal of Animal Science* 75(11), 2899-2904.

Hasan, S., Orro, T., Valros, A., Junnikkala, S., Peltoniemi, O. and Oliviero, C. 2019. Factors affecting sow colostrum yield and composition, and their impact on piglet growth and health. *Livestock Science* 227, 60-67.

Heidinger, B., Stinglmayr, J., Maschat, K., Oberer, M., Kuchling, S. and Baumgartner, J. 2018. Summary of the Austrian project "pro-SAU": evaluation of novel farrowing systems with possibility for the sow to move 2410:Eval. AFB Supplement zum Abschlussbericht des Projekts Pro-SAU.

Herpin, P., Damon, M. and Le Dividich, J. 2002. Development of thermoregulation and neonatal survival in pigs. *Livestock Production Science* 78(1), 25-45.

Hessel, E. F., Reiners, K. and Van den Weghe, H. F. A. 2006. Socializing piglets before weaning: effects on behavior of lactating sows, pre-and postweaning behavior, and performance of piglets. *Journal of Animal Science* 84(10), 2847-2855.

Homedes, J., Salichs, M., Sabaté, D., Sust, M. and Fabre, R. 2014. Effect of ketoprofen on pre-weaning piglet mortality on commercial farms. *The Veterinary Journal* 201(3), 435-437.

Honeyman, M. S., Roush, W. B. and Penner, A. D. 1998. Pig crushing mortality by hut type in outdoor farrowing. Annual Progress Report, Iowa State University, Ames, pp. 16-17.

Horrell, I. and Bennett, J. 1981. Disruption of teat preferences and retardation of growth following cross-fostering of 1-week-old pigs. *Animal Science* 33(1), 99-106.

Hötzel, M. J., Pinheiro Machado F°, L. C., Wolf, F. M. and Costa, O. A. D. 2004. Behaviour of sows and piglets reared in intensive outdoor or indoor systems. *Applied Animal Behaviour Science* 86(1-2), 27-39.

Hunt, K. and Petchey, A. M. 1987. A study of the environmental preferences of sows around farrowing. *Farm Building Progress* 89, 11-14.

Ison, S. H., Wood, C. M. and Baxter, E. M. 2015. Behaviour of pre-pubertal gilts and its relationship to farrowing behaviour in conventional farrowing crates and loose-housed pens. *Applied Animal Behaviour Science* 170, 26-33.

Ison, S. H., Clutton, R. E., Di Giminiani, P. and Rutherford, K. M. 2016. A review of pain assessment in pigs. *Frontiers in Veterinary Science* 3, 108.

Ison, S. H., Jarvis, S., Hall, S. A., Ashworth, C. J. and Rutherford, K. M. D. 2018. Periparturient behavior and physiology: further insight into the farrowing process for primiparous and multiparous sows. *Frontiers in Veterinary Science* 5, 122.

Jarrett, S. and Ashworth, C. J. 2018. The role of dietary fibre in pig production, with a particular emphasis on reproduction. *Journal of Animal Science and Biotechnology* 9(1), 1-11.

Jarvis, S., Lawrence, A. B., McLean, K. A., Deans, L. A., Chirnside, J. and Calvert, S. K. 1997. The effect of environment on behavioural activity, ACTH, b-endorphin and cortisol in pre-parturient gilts. *Animal Science* 65(3), 465-472.

Jarvis, S., Lawrence, A. B., McLean, K. A., Chirnside, J., Deans, L. A. and Calvert, S. K. 1998. The effect of environment on plasma cortisol and b-endorphin in the parturient pig and the involvement of endogenous opioids. *Animal Reproduction Science* 52(2), 139-151.

Jarvis, S., McLean, K. A., Calvert, S. K., Deans, L. A., Chirnside, J. and Lawrence, A. B. 1999. The responsiveness of sows to their piglets in relation to the length of parturition and the involvement of endogenous opioids. *Applied Animal Behaviour Science* 63(3), 195-207.

Jarvis, S., van der Vegt, B. J., Lawrence, A. B., McLean, K. A., Deans, L. A., Chirnside, J. and Calvert, S. K. 2001. The effect of parity and environmental restriction on behavioural and physiological responses of pre-parturient pigs. *Applied Animal Behaviour Science* 71(3), 203-216.

Jarvis, S., Calvert, S. K., Stevenson, J., van Leeuwen, N. and Lawrence, A. B. 2002. Pituitary-adrenal activation in pre-parturient pigs (Sus scrofa) is associated with behavioural restriction due to lack of space rather than nesting substrate. *Animal Welfare* 11, 371-384.

Jarvis, S., Reed, B. T., Lawrence, A. B., Calvert, S. K. and Stevenson, J. 2004. Peri-natal environmental effects on maternal behaviour, pituitary and adrenal activation, and the progress of parturition in the primiparous sow. *Animal Welfare* 13(2), 171-181.

Jarvis, S., D'Eath, R. B., Robson, S. K. and Lawrence, A. B. 2006. The effect of confinement during lactation on the hypothalamic-pituitary-adrenal axis and behaviour of primiparous sows. *Physiology and Behavior* 87(2), 345-352.

Jensen, P. 1986. Observations on the maternal behaviour of free-ranging domestic pigs. *Applied Animal Behaviour Science* 16(2), 131-142.

Jensen, P. 1993. Nest building in domestic sows: the role of external stimuli. *Animal Behaviour* 45(2), 351-358.

Jensen, P. 2002. Behaviour of pigs. In: P Jensen (Eds), *The Ethology of Domestic Animals*, pp. 159-172. CABI Publishing, Wallingford, UK.

Jensen, P. and Toates, F. M. 1993. Who needs 'behavioural needs'? Motivational aspects of the needs of animals. *Applied Animal Behaviour Science* 37(2), 161-181.

Jones, G. M., Edwards, S. A., Sinclair, A. G., Gebbie, F. E., Rooke, J. A., Jagger, S. and Hoste, S. 2002. The effect of maize starch or soya-bean oil as energy sources in lactation on sow and piglet performance in association with sow metabolic state around peak lactation. *Animal Science* 75(1), 57-66.

King, R. L., Baxter, E. M., Matheson, S. M. and Edwards, S. A. 2019. Temporary crate opening procedure affects immediate post-opening piglet mortality and sow behaviour. *Animal: An International Journal of Animal Bioscience* 13(1), 189-197.

Kobek-Kjeldager, C., Moustsen, V. A., Theil, P. K. and Pedersen, L. J. 2020. Effect of litter size, milk replacer and housing on production results of hyper-prolific sows. *Animal* 14(4), 824-833.

Kyriazakis, I. and Edwards, S. A. 1986. The effects of split suckling on behaviour and performance of piglets. *Applied Animal Behaviour Science* 16(1), 92.

Ladewig, J., Kloeppel, P. and Kallweit, E. 1984. A case of "reversed cannibalism" the piglets damaging the sow [vulvar lesions]. *Annales de Recherches Veterinaires (France)*. Annals of veterinary research, 15(2), pp.275-277.

Lawrence, A. B., McLean, K. A., Jarvis, S., Gilbert, C. L. and Petherick, J. C. 1997. Stress and parturition in the pig. *Reproduction in Domestic Animals* 32(5), 231-236.

Lawrence, A. B., Petherick, J. C., McLean, K. A., Deans, L. A., Chirnside, J., Gaughan, A., Clutton, E. and Terlouw, E. M. C. 1994. The effect of environment on behaviour, plasma cortisol and prolactin in parturient sows. *Applied Animal Behaviour Science* 39(3-4), 313-330.

Lawrence, A. B., Petherick, J. C., Mclean, K. and Gilbert, C. 1992. Mediation of stress-induced inhibition of oxytocin in farrowing sows by endogenous opioids. *Proceedings of the British Society of Animal Production* 1992, 56.

Leeb, B., Leeb, C., Troxler, J. and Schuh, M. 2001. Skin lesions and callosities in group-housed pregnant sows: animal-related welfare indicators. *Acta Agriculturae Scandinavica, Section A – Animal Science* 51(S30), 82-87.

Lensink, B. J., Ofner-Schröck, E., Ventorp, M., Zappavigna, P., Flaba, J., Georg, H. and Bizeray-Filoche, D. 2013. In Andres Aland, Thomas Banhazi (Eds) Lying and walking surfaces for cattle, pigs and poultry and their impact on health, behaviour and performance. In: *Livestock Housing: Modern Management to Ensure Optimal Health and Welfare of Farm Animals*, pp. 2509-2514. Wageningen Academic Publishers, Wageningen, The Netherlands.

Lewis, E., Boyle, L. A., Brophy, P., O'Doherty, J. V. and Lynch, P. B. 2005. The effect of two piglet teeth resection procedures on the welfare of sows in farrowing crates. Part 2. *Applied Animal Behaviour Science* 90(3-4), 251-264.

Lohmeier, R. Y., Grimberg-Henrici, C. G. E., Büttner, K., Burfeind, O. and Krieter, J. 2020. Farrowing pens used with and without short-term fixation impact on reproductive traits of sows. *Livestock Science* 231, 103889.

Loisel, F., Farmer, C., Ramaekers, P. and Quesnel, H. 2013. Effects of high fiber intake during late pregnancy on sow physiology, colostrum production, and piglet performance. *Journal of Animal Science* 91(11), 5269-5279.

Mainau, E., Ruiz-de-la-Torre, J. L., Dalmau, A., Salleras, J. M. and Manteca, X. 2012. Effects of meloxicam (Metacam®) on post-farrowing sow behaviour and piglet performance. *Animal: An International Journal of Animal Bioscience* 6(3), 494-501.

Mainau, E., Temple, D. and Manteca, X. 2016. Experimental study on the effect of oral meloxicam administration in sows on pre-weaning mortality and growth and immunoglobulin G transfer to piglets. *Preventive Veterinary Medicine* 126, 48-53.

Malmkvist, J., Pedersen, L. J., Kammersgaard, T. S. and Jørgensen, E. 2012. Influence of thermal environment on sows around farrowing and during the lactation period. *Journal of Animal Science* 90(9), 3186-3199.

Marchant, J. N., Broom, D. M. and Corning, S. 2001. The influence of sow behaviour on piglet mortality due to crushing in an open farrowing system. *Animal Science* 72(1), 19-28.

Martin, J. E., Ison, S. H. and Baxter, E. M. 2015. The influence of neonatal environment on piglet play behaviour and post-weaning social and cognitive development. *Applied Animal Behaviour Science* 163, 69-79.

Mattsson, B., Susic, Z. and Lundeheim, N. 2004. Time for care and comfort. In Proceedings of 18th IPVS Congress, Hamburg, Germany, p. 800.

Middelkoop, A., Choudhury, R., Gerrits, W. J. J., Kemp, B., Kleerebezem, M. and Bolhuis, J. E. 2018. Dietary diversity affects feeding behaviour of suckling piglets. *Applied Animal Behaviour Science* 205, 151-158.

Middelkoop, A., Costermans, N., Kemp, B. and Bolhuis, J. E. 2019. Feed intake of the sow and playful creep feeding of piglets influence piglet behaviour and performance before and after weaning. *Scientific Reports* 9(1), 16140.

Morgan, T., Pluske, J., Miller, D., Collins, T., Barnes, A. L., Wemelsfelder, F. and Fleming, P. A. 2014. Socialising piglets in lactation positively affects their post-weaning behaviour. *Applied Animal Behaviour Science* 158, 23-33.

Mount, L. E. 1967. The heat loss from new born pigs to the floor. *Research in Veterinary Science* 8(2), 175-186.

Mount, L. E. 1968. *The Climatic Physiology of the Pig*. Edward Arnold, London, UK.

Moustsen, V. A. and Poulsen, H. L. 2004. Anbefalinger vedr. dimensioner på farebox og kassesti. *Landsudvalget for Svin, Danske Slagterier, Notat* nr. 414.

Moustsen, V. and Jensen, T. 2008. *Inventar til forbedring af hygiejne i stier til løsgående farende og diegievende søer*. Notat, Nr. 809, Dansk Svineproduktion.

Moustsen, V. A., Pedersen, L. J. and Jensen, T. 2007. Afprøvning af stikoncepter til løsgående farende og diegivende søer. Meddelelse (805): 1-25. Dansk Svineproduktion.

Moustsen, V. A., Lahrmann, H. P. and D'Eath, R. B. 2011. Relationship between size and age of modern hyper-prolific crossbred sows. *Livestock Science* 141(2-3), 272-275.

Moustsen, V. A., Hales, J., Lahrmann, H. P., Weber, P. M. and Hansen, C. F. 2013. Confinement of lactating sows in crates for 4 days after farrowing reduces piglet mortality. *Animal* 7(4), 648-654.

Mouttotou, N. and Green, L. E. 1999. Incidence of foot and skin lesions in nursing piglets and their association with behavioural activities. *Veterinary Record* 145(6), 160-165.

Muns, R., Silva, C., Manteca, X. and Gasa, J. 2014. Effect of cross-fostering and oral supplementation with colostrums on performance of newborn piglets. *Journal of Animal Science* 92(3), 1193-1199.

Muns, R., Manteca, X. and Gasa, J. 2015. Effect of different management techniques to enhance colostrum intake on piglets' growth and mortality. *Animal Welfare* 24(2), 185-192.

Muns, R., Malmkvist, J., Larsen, M. L. V., Sørensen, D. and Pedersen, L. J. 2016. High environmental temperature around farrowing induced heat stress in crated sows. *Journal of Animal Science* 94(1), 377-384.

Muns, R., Nuntapaitoon, M. and Tummaruk, P. 2017. Effect of oral supplementation with different energy boosters in newborn piglets on pre-weaning mortality, growth and serological levels of IGF-I and IgG. *Journal of Animal Science* 95(1), 353-360.

Nissen, S., Faidley, T. D., Zimmerman, D. R., Izard, R. and Fisher, C. T. 1994. Colostral milk fat percentage and pig performance are enhanced by feeding the leucine metabolite β-hydroxy-β-methyl butyrate to sows. *Journal of Animal Science* 72(9), 2331-2337.

Nowland, T. L., van Wettere, W. H. E. J. and Plush, K. J. 2019. Allowing sows to farrow unconfined has positive implications for sow and piglet welfare. *Applied Animal Behaviour Science* 221, 104872.

Ocepek, M. and Andersen, I. L. 2017. What makes a good mother? Maternal behavioural traits important for piglet survival. *Applied Animal Behaviour Science* 193, 29-36.

Ocepek, M., Goold, C. M., Busančić, M. and Aarnink, A. J. A. 2018. Drinker position influences the cleanness of the lying area of pigs in a welfare-friendly housing facility. *Applied Animal Behaviour Science* 198, 44-51.

Oliviero, C., Heinonen, M., Valros, A. and Peltoniemi, O. 2010. Environmental and sow-related factors affecting the duration of farrowing. *Animal Reproduction Science* 119(1-2), 85-91.

O'Mahony, S. M., Marchesi, J. R., Scully, P., Codling, C., Ceolho, A. M., Quigley, E. M., Cryan, J. F. and Dinan, T. G. 2009. Early life stress alters behavior, immunity, and microbiota in rats: implications for irritable bowel syndrome and psychiatric illnesses. *Biological Psychiatry* 65(3), 263-267.

Oostindjer, M., Bolhuis, J. E., Mendl, M., Held, S., Gerrits, W., Van den Brand, H. and Kemp, B. 2010. Effects of environmental enrichment and loose housing of lactating sows on piglet performance before and after weaning. *Journal of Animal Science* 88(11), 3554-3562.

Oostindjer, M., van den Brand, H., Kemp, B. and Bolhuis, J. E. 2011a. Effects of environmental enrichment and loose housing of lactating sows on piglet behaviour before and after weaning. *Applied Animal Behaviour Science* 134(1-2), 31-41.

Oostindjer, M., Bolhuis, J. E., Mendl, M., Held, S., van den Brand, H. and Kemp, B. 2011b. Learning how to eat like a pig: effectiveness of mechanisms for vertical social learning in piglets. *Animal Behaviour* 82(3), 503-511.

Oostindjer, M., Bolhuis, J. E., Simon, K., van den Brand, H. and Kemp, B. 2011c. Perinatal flavour learning and adaptation to being weaned: all the pig needs is smell. *PLoS ONE* 6(10), e25318.

Oostindjer, M., Kemp, B., van den Brand, H. and Bolhuis, J. E. 2014. Facilitating 'learning from mom how to eat like a pig' to improve welfare of piglets around weaning. *Applied Animal Behaviour Science* 160, 19-30.

Pajor, E. A. 1998. Parent-offspring conflict and its implications for maternal housing systems in domestic pigs. Doctoral dissertation, McGill University Libraries.

Pajor, E. A., Weary, D. M., Fraser, D. and Kramer, D. L. 1999. Alternative housing for sows and litters: 1. Effects of sow-controlled housing on responses to weaning. *Applied Animal Behaviour Science* 65(2), 105-121.

Patil, Y., Gooneratne, R. and Ju, X. H. 2020. Interactions between host and gut microbiota in domestic pigs: a review. *Gut Microbes*, 11(3), 310-334.

Pedersen, L. J., Jørgensen, E., Heiskanen, T. and Damm, B. I. 2006. Early piglet mortality in loose-housed sows related to sow and piglet behaviour and to the progress of parturition. *Applied Animal Behaviour Science* 96(3-4), 215-232.

Pedersen, L. J., Malmkvist, J. and Andersen, H. M. L. 2013. Housing of sows during farrowing: a review on pen design, welfare and productivity. In: Andres Aland and Thomas Banhazi (Eds), *Livestock Housing: Modern Management to Ensure Optimal Health and Welfare of Farm Animals,* pp. 285-297. Wageningen Academic Publishers. Wageningen, The Netherlands.

Pedersen, M. L., Moustsen, V. A., Nielsen, M. B. F. and Kristensen, A. R. 2011. Improved udder access prolongs duration of milk letdown and increases piglet weight gain. *Livestock Science* 140(1-3), 253-261.

Pedersen, T. F., Van Vliet, S., Bruun, T. S. and Theil, P. K. 2019. Feeding sows during the transition period—is a gestation diet, a simple transition diet, or a lactation diet the best choice? *Translational Animal Science* 4(1), 34-48.

Peng, X., Yan, C., Hu, L., Liu, Y., Xu, Q., Wang, R., Qin, L., Wu, C., Fang, Z., Lin, Y., Xu, S., Feng, B., Zhuo, Y., Li, J., Wu and Che, L. 2019. Effects of fat supplementation during gestation on reproductive performance, milk composition of sows and intestinal development of their offspring. *Animals* 9(4), 125.

Petherick, J. C. 1983. A biological basis for the design of space in livestock housing. In: Baxter, S. H., Baxter, M. R. and MacCormack, J. A. S. C. (Eds), *Farm Animal Housing and Welfare*, pp. 103-120. Martinus Nijoff Publisher, Boston, MA.

Phillips, P. A., Fraser, D. and Pawluczuk, B. 2000. Floor temperature preference of sows at farrowing. *Applied Animal Behaviour Science* 67(1-2), 59-65.

Plush, K., McKenny, L., Nowland, T., van Wettere, W. and Terry, R. 2019. *Reducing Sow Stress around Farrowing*. Report prepared for the Co-operative Research Centre for High Integrity Australian Pork. *(Pork CRC) project # 1C-114)*.

Power, M. L. and Schulkin, J. 2013. Maternal regulation of offspring development in mammals is an ancient adaptation tied to lactation. *Applied and Translational Genomics* 2, 55-63.

Price, E. O., Hutson, G. D., Price, M. I. and Borgwardt, R. 1994. Fostering in swine as affected by age of offspring. *Journal of Animal Science* 72(7), 1697-1701.

Puppe, B. and Tuchscherer, A. 1999. Developmental and territorial aspects of suckling behaviour in the domestic pig (Sus scrofa f. domestica). *Journal of Zoology* 249(3), 307-313.

Quesnel, H., Farmer, C. and Devillers, N. 2012. Colostrum intake: influence on piglet performance and factors of variation. *Livestock Science* 146(2-3), 105-114.

Quiniou, N. and Noblet, J. 1999. Influence of high ambient temperatures on performance of multiparous lactating sows. *Journal of Animal Science* 77(8), 2124-2134.

Randall, J. M., Armsby, A. W. and Sharp, J. R. 1983. Cooling gradients across pens in a finishing piggery: II. *Journal of Agricultural Engineering Research* 28(3), 247-259.

Rantzer, D. and Svendsen, J. 2001. Slatted versus solid floors in the dung area of farrowing pens: effects on hygiene and pig performance, birth to weaning. *Acta Agriculturae Scandinavica, Section A - Animal Science* 51(3), 167-174.

Renaudeau, D. and Noblet, J. 2001. Effects of exposure to high ambient temperature and dietary protein level on sow milk production and performance of piglets. *Journal of Animal Science* 79(6), 1540-1548.

Rioja-Lang, F. C., Seddon, Y. M. and Brown, J. A. 2018. Shoulder lesions in sows: a review of their causes, prevention, and treatment. *Journal of Swine Health and Production* 26(2), 101-107.

Robert, S. and Martineau, G. P. 2001. Effects of repeated crossfosterings on preweaning behavior and growth performance of piglets and on maternal behavior of sows. *Journal of Animal Science* 79(1), 88-93.

Roehe, R. 1999. Genetic determination of individual birth weight and its association with sow productivity traits using Bayesian analyses. *Journal of Animal Science* 77(2), 330-343.

Rolandsdotter, E., Westin, R. and Algers, B. 2009. Maximum lying bout duration affects the occurrence of shoulder lesions in sows. *Acta Veterinaria Scandinavica* 51(1), 44.

Rooke, J. A., Sinclair, A. G., Edwards, S. A., Cordoba, R., Pkiyach, S., Penny, P. C., Penny, P., Finch, A. M. and Horgan, G. W. 2001. The effect of feeding salmon oil to sows throughout pregnancy on pre-weaning mortality of piglets. *Animal Science* 73(3), 489-500.

Rosvold, E. M. and Andersen, I. L. 2019. Straw vs. peat as nest-building material-the impact on farrowing duration and piglet mortality in loose-housed sows. *Livestock Science* 229, 203-209.

Rosvold, E. M., Kielland, C., Ocepek, M., Framstad, T., Fredriksen, B., Andersen-Ranberg, I., Næss, G. and Andersen, I. L. 2017. Management routines influencing piglet survival in loose-housed sow herds. *Livestock Science* 196, 1-6.

Rosvold, E. M., Newberry, R. C., Framstad, T. and Andersen, I. L. 2018. Nest-building behaviour and activity budgets of sows provided with different materials. *Applied Animal Behaviour Science* 200, 36-44.

Ryan, E. B., Fraser, D. and Weary, D. M. 2015. Public attitudes to housing systems for pregnant pigs. *PLoS ONE* 10(11), e0141878.

Schmitt, O., Baxter, E. M., Lawlor, P. G., Boyle, L. A. and O'Driscoll, K. 2019a. A single dose of fat-based energy supplement to light birth weight pigs shortly after birth does not increase their survival and growth. *Animals* 9(5), 227.

Schmitt, O., Baxter, E. M., Boyle, L. A. and O'Driscoll, K. 2019b. Nurse sow strategies in the domestic pig: I. Consequences for selected measures of sow welfare. *Animal* 13(3), 580-589.

Schmitt, O., Baxter, E. M., Boyle, L. A. and O'Driscoll, K. 2019c. Nurse sow strategies in the domestic pig: II. Consequences for piglet growth, suckling behaviour and sow nursing behaviour. *Animal* 13(3), 590-599.

Schmitt, O., O'Driscoll, K., Boyle, L. A. and Baxter, E. M. 2019d. Artificial rearing affects piglets pre-weaning behaviour, welfare and growth performance. *Applied Animal Behaviour Science* 210, 16-25.

Skok, J. and Škorjanc, D. 2014. Group suckling cohesion as a prelude to the formation of teat order in piglets. *Applied Animal Behaviour Science* 154, 15-21.

Sørensen, J. T. and Fraser, D. 2010. On-farm welfare assessment for regulatory purposes: issues and possible solutions. *Livestock Science* 131(1), 1-7.

Sørensen, J. T., Rousing, T., Kudahl, A. B., Hansted, H. J. and Pedersen, L. J. 2016. Do nurse sows and foster litters have impaired animal welfare? Results from a cross-sectional study in sow herds. *Animal* 10(4), 681–686.

Strathe, A. V., Bruun, T. S. and Hansen, C. F. 2017. Sows with high milk production had both a high feed intake and high body mobilization. *Animal* 11(11), 1913–1921.

Straw, B. E., Dewey, C. E. and Bürgi, E. J. 1998. Patterns of crossfostering and piglet mortality on commercial US and Canadian swine farms. *Preventive Veterinary Medicine* 33(1–4), 83–89.

Swan, K. M., Peltoniemi, O. A. T., Munsterhjelm, C. and Valros, A. 2018. Comparison of nest-building materials in farrowing crates. *Applied Animal Behaviour Science* 203, 1–10.

Telkänranta, H. and Edwards, S. A. 2018. Lifetime consequences of the early physical and social environment of piglets. In: Špinka, M. (Ed.), *Advances in Pig Welfare* pp. 101–136. Woodhead Publishing, Cambridge, UK.

Theil, P. K., Lauridsen, C. and Quesnel, H. 2014. Neonatal piglet survival: impact of sow nutrition around parturition on fetal glycogen deposition and production and composition of colostrum and transient milk. *Animal: An International Journal of Animal Bioscience* 8(7), 1021–1030.

Thodberg, K., Jensen, K. H., Herskin, M. S. and Jorgensen, E. 1999. Influence of environmental stimuli on nest building and farrowing behaviour in domestic sows. *Applied Animal Behaviour Science* 63(2), 131–144.

Thøgersen, J. and Zhou, Y. 2012. Chinese consumers' adoption of a "green" innovation–the case of organic food. *Journal of Marketing Management* 28(3–4), 313–333.

Thomsson, O., Sjunnesson, Y., Magnusson, U., Eliasson-Selling, L., Wallenbeck, A. and Bergqvist, A. S. 2016. Consequences for piglet performance of group housing lactating sows at one, two, or three weeks post-farrowing. *PLoS ONE* 11(6), e0156581.

Thorup, F., Wedel-Müller, R. L., Hansen, C. F., Kanitz, E. and Tuchscherer, M. 2015. Neonatal mortality in piglets is more due to lack of energy than lack of immunoglobulins. In: *International Conference on Pig Welfare: Improving Pig Welfare-What Are the Ways Forward?*, p. 84. Wageningen Academic Publishers; Copenhagen, Denmark

Tuchscherer, M., Puppe, B., Tuchscherer, A. and Tiemann, U. 2000. Early identification of neonates at risk: traits of newborn piglets with respect to survival. *Theriogenology* 54(3), 371–388.

Turner, S. P., Camerlink, I., Baxter, E. M., D'Eath, R. B., Desire, S. and Roehe, R. 2018. Breeding for pig welfare: opportunities and challenges. In: Špinka, M. (Ed.), *Advances in Pig Welfare*, pp. 399–414. Woodhead Publishing, Cambridge, UK.

Tybirk, P., Sloth, N. M. and Jorgensen, L. 2012. *Danish Nutrient Requirement Standards (in Danish: Normer for Naringsstoffer)* (17th rev. edn.). SEGES Pig Research Centre, Axelborg, Denmark.

Tybirk, P., Sloth, N. M., Sonderby, T. B.. and Kjeldsen, N. 2015. *Danish Nutrient Requirement Standards (in Danish Normer for Naringsstoffer)* (22th rev. edn.). SEGES Pig Research Centre, Axelborg, Denmark.

van Beirendonck, S., Schroijen, B., Bulens, A., Van Thielen, J. and Driessen, B. 2015. A solution for high production numbers in farrowing units? In: *International Conference on Pig Welfare: Improving Pig Welfare-What Are the Ways Forward*, pp. 85–85. Wageningen Academic Publishers, Copenhagen, Denmark.

van de Weerd, H. and Ison, S. 2019. Providing effective environmental enrichment to pigs: how far have we come? *Animals* 9(5), 254.

van Dixhoorn, I. D., Reimert, I., Middelkoop, J., Bolhuis, J. E., Wisselink, H. J., Koerkamp, P. W. G., Kemp, B. and Stockhofe-Zurwieden, N. 2016. Enriched housing reduces disease susceptibility to co-infection with porcine reproductive and respiratory virus (PRRSV) and Actinobacillus pleuropneumoniae (A. pleuropneumoniae) in young pigs. *PLoS ONE* 11(9), e0161832.

van Nieuwamerongen, S. E., Bolhuis, J. E., Van der Peet-Schwering, C. M. C. and Soede, N. M. 2014. A review of sow and piglet behaviour and performance in group housing systems for lactating sows. *Animal* 8(3), 448-460.

van Nieuwamerongen, S. E., Soede, N. M., van der Peet-Schwering, C. M. C., Kemp, B. and Bolhuis, J. E. 2015. Development of piglets raised in a new multi-litter housing system vs. conventional single-litter housing until 9 weeks of age. *Journal of Animal Science* 93(11), 5442-5454.

Verbeke, W. 2009. Stakeholder, citizen and consumer interests in farm animal welfare. *Animal Welfare* 18(4), 325-333.

Verdon, M., Morrison, R. S. and Rault, J. L. 2019. Sow and piglet behaviour in group lactation housing from 7 or 14 days post-partum. *Applied Animal Behaviour Science* 214, 25-33.

Wathes, C. and Whittemore, C. T. 2006. Environmental management of pigs. In: Kyriazakis, I. and Whittemore, C. T. (Eds), *Whittemore's Science and Practice of Pig Production*, pp. 533-592. Blackwell Publishing, Oxford, UK.

Weary, D. M., Pajor, E. A., Bonenfant, M., Fraser, D. and Kramer, D. L. 2002. Alternative housing for sows and litters. Part 4. *Applied Animal Behaviour Science* 76(4), 279-290.

Weary, D. M., Jasper, J. and Hötzel, M. J. 2008. Understanding weaning distress. *Applied Animal Behaviour Science* 110(1-2), 24-41.

Westin, R., Holmgren, N., Hultgren, J. and Algers, B. 2014. Large quantities of straw at farrowing prevents bruising and increases weight gain in piglets. *Preventive Veterinary Medicine* 115(3-4), 181-190.

Widowski, T. M. and Curtis, S. E. 1990. The influence of straw, cloth tassel, or both on the prepartum behaviour of sows. *Applied Animal Behaviour Science* 27(1-2), 53-71.

Wiegand, R. M., Gonyou, H. W. and Curtis, S. E. 1994. Pen shape and size: effects on pig behavior and performance. *Applied Animal Behaviour Science* 39(1), 49-61.

Williams, A. M., Safranski, T. J., Spiers, D. E., Eichen, P. A., Coate, E. A. and Lucy, M. C. 2013. Effects of a controlled heat stress during late gestation, lactation, and after weaning on thermoregulation, metabolism, and reproduction of primiparous sows. *Journal of Animal Science* 91(6), 2700-2714.

Wischner, D., Kemper, N. and Krieter, J. 2009. Nest-building behaviour in sows and consequences for pig husbandry. *Livestock Science* 124(1-3), 1-8.

Wu, G., Bazer, F. W., Johnson, G. A., Knabe, D. A., Burghardt, R. C., Spencer, T. E., Li, X. L. and Wang, J. J. 2011. Triennial Growth Symposium: important roles for L-glutamine in swine nutrition and production *Journal of Animal Science* 89(7), 2017-2030.

You, X., Li, Y., Zhang, M., Yan, H. and Zhao, R. 2014. A survey of Chinese citizens' perceptions on farm animal welfare. *PLoS ONE* 9(10), e109177.

Yun, J., Swan, K. M., Vienola, K., Kim, Y. Y., Oliviero, C., Peltoniemi, O. A. T. and Valros, A. 2014. Farrowing environment has an impact on sow metabolic status and piglet colostrum intake in early lactation. *Livestock Science* 163, 120-125.

Yun, J. and Valros, A. 2015. Benefits of prepartum nest-building behaviour on parturition and lactation in sows—a review. *Asian-Australasian Journal of Animal Sciences* 28(11), 1519-1524.

Yunes, M. C., von Keyserlingk, M. A. G. and Hötzel, M. J. 2017. Brazilian citizens' opinions and attitudes about farm animal production systems. *Animals* 7(10), 75.

Zanella, A. J. and Zanella, E. L. 1993. Nesting material used by free-range sows in Brazil. In: Nichelmann, M., Wierenga, H. K. and Braun, S. (Eds), *Proceedings of the 3rd Joint Meeting of the International Congress on Applied Ethology*, p. 411. Humboldt-Universitaet, Berlin.

Zoric, M., Nilsson, E., Lundeheim, N. and Wallgren, P. 2009. Incidence of lameness and abrasions in piglets in identical farrowing pens with four different types of floor. *Acta Veterinaria Scandinavica* 51(1), 23.

Zurbrigg, K. 2006. Sow shoulder lesions: risk factors and treatment effects on an Ontario farm. *Journal of Animal Science* 84(9), 2509–2514.

Chapter 4

Welfare of gilts and pregnant sows

Sandra Edwards, Newcastle University, UK

1 Introduction

The wild ancestors and wild relatives of the modern domestic sow live in small stable family groups. Many of the behaviour patterns that they show have been conserved in modern sows which, when placed back into a semi-natural environment, spend 0.6 of the daylight hours grazing and rooting (Stolba and Wood-Gush, 1989). In addition to feeding, drinking and excretory behaviours, maintenance behaviours include those with thermoregulatory function, such as nest building and wallowing, and others with skin care functions, such as rubbing. In this complex environment, exploratory behaviours, including locomotion, orientation to stimuli, nosing and manipulation of objects, also occupy significant periods of time (0.1–0.2 of the day). The stable social situation is reflected in the spatial association of stable subgroups, low incidence of agonistic behaviours and social facilitation of behaviour within the group. In intensive farmed conditions, circumstances for the pregnant sow are greatly changed. Food is provided in a concentrated form, and foraging behaviours are no longer functional in increasing energy and nutrient supply. Long-term stable group structure is disrupted, space allowance is greatly reduced and

http://dx.doi.org/10.19103/AS.2017.0013.22

the environment is frequently utilitarian and barren. This situation restricts expression of the behaviour patterns developed by sows in their evolutionary history and may give rise to behaviours considered abnormal and indicative of reduced welfare.

Traditional systems, in which small groups of pregnant sows were housed in outdoor paddocks or in covered straw yards, fell into disfavour as herd sizes increased, and it became more difficult to manage them. The 1960s saw the large-scale development and adoption of individual gestation stall and tether housing systems for sows, and these rapidly became the norm in many pig-producing countries. Such systems offered the advantages of low space requirement and ease of management. Sows could no longer fight and individual feeding could ensure that the nutritional needs of all animals were precisely met without competition. With the small space allowances and enclosed buildings associated with individual housing, automated air temperature control was possible, but provision of bedding and daily cleaning-out presented a difficult and laborious manual task. In consequence, this was automated in many buildings by using fully or partly slatted floors, through which all excreta passed for storage away from the animals as a slurry. The slurry could then be mechanically removed from the building at any convenient time. The resulting form of housing offered a relatively low cost, simply managed system for pig production enterprises under all farming conditions (Fig. 1). However, the large-scale commercial use of such systems became a focus for public concern because of the associated indications of welfare impairment for the animals.

Figure 1 Gestation stall housing which allows protected individual feeding, but restricts space and foraging opportunity. Photo courtesy of Prairie Swine Centre, Canada.

2 Welfare issues of individual confinement systems

Following concerns publicised in widely read articles and books, such as Ruth Harrison's *Animal Machines* published in 1964, scientific studies of the consequences of different housing systems during gestation multiplied during the 1980s. Comparison of sows housed in stalls with those housed in small groups of 4–5 animals in either cubicles (pens in which animals can move freely between stalls used for feeding/resting and a communal dunging/exercise area) or in larger pens with separate lying, activity and feeding areas showed that confined animals spent more time lying down, had shorter activity periods and spent significant periods of time chewing on the bars of the pen (Jensen, 1980a). When straw was provided, sows housed in stalls ate it all quickly, whereas those in loose housing showed a variety of different behaviour patterns associated with nest building and exploration. These early behavioural indications of problems with confined housing were followed by more detailed experimental studies comparing the development of behaviour of groups of animals from a common background when they were put into different housing systems. Similar results were obtained in comparisons of stalls and group housing (Svendsen and Bengtsson, 1983) or tethers and loose housing (Vestergaard and Hansen, 1984). It was again shown that confined animals, in either system, were less active, showed less manipulation of straw and spent more time bar and chain biting and drinking. Behavioural differences persisted when the animals were taken to farrowing accommodation, with those previously housed in stalls or tethers being more restless at farrowing and having longer delivery intervals. Despite being unable to physically interact, tethered animals also showed more aggressive behaviours directed at neighbouring animals.

To assess in more detail the significance of the differences in behaviour in different housing systems for animal welfare, a series of experiments was carried out in Australia. A preliminary study compared non-pregnant gilts in five different housing systems: individually tethered, housed in pairs, indoor group pens, outdoor concrete yards with simple shelter and outdoor paddocks (Barnett et al., 1984). As in previous studies, outdoor groups were more active, whilst indoor groups, especially those in tethers and pairs, spent more time sham chewing, bar biting and drinking. Assessment of animal welfare by measuring the diurnal levels of blood corticosteroids showed that animals of the indoor group pens had lowest levels, whilst those housed in pairs in limited space appeared to exhibit a chronic stress response. This was reflected in higher cortisol levels, a disrupted diurnal pattern of blood cortisol and slower response to and recovery from a transport stressor. A subsequent study with pregnant pigs compared tethers, stalls, indoor groups and paddocks (Barnett et al., 1985). Again outdoor pigs were found to be more active, spending about 0.2 of the daylight hours in rooting and grazing. Although animals in stalls showed increased oral/nasal behaviours, it was those in tethers which showed lower activity and evidence of a chronic physiological stress response. The extent of the increase in corticosteroids in tethered animals was sufficient to induce metabolic changes, reflected in higher blood glucose and urea levels (Barnett et al., 1985) and reduced responsiveness of the immune system to an external challenge (Barnett et al., 1987a).

In a later study (Barnett et al., 1987a) it was noted that sows in tethers, whilst performing less exploratory behaviour than those in groups, actually had more aggressive interactions with neighbours and more retaliatory behaviour in such interactions. This has led to an investigation of the possibility that reduced welfare of tethered animals was at least partially due to social stress arising from being in enforced close proximity to a neighbouring animal and might be improved by altering tether design. When the partition between adjacent tethered sows was changed from one of vertical bars to a wire mesh, the total number of interactions was decreased, aggressive interactions were virtually eliminated and corticosteroid levels were similar to those of stall or group-housed animals (Barnett et al., 1987b). Barnett et al. (1989) therefore suggested that stall housing was less stressful than tethers because the increased opportunity for back and forward movement reduced the amount of head-to-head contact between pigs. An effect of stall design on the nature of interactions and stress responses was subsequently found (Barnett et al., 1991), but this was in contrast to the previous theory. Animals in stalls with horizontal bars spent less time concurrently in front of the stall with a neighbour but, despite lower levels of aggression and head-to-head contact, had higher cortisol levels than those in stalls with vertical bars. Thus, the extent to which the physiological indicators of stress associated with confinement housing can be explained by social factors is still incomplete.

3 Nature and significance of stereotyped behaviour in gestating sows

As described previously, the majority of studies have shown a marked increase in abnormal oral behaviours in confined sows. In many cases such behaviours develop into stereotypies – repetitive behaviours which are fixed in pattern of performance and have no obvious function. The range and frequency of abnormal oral/nasal behaviours were described by Cronin and Wiepkema (1984) who studied in detail 36 neck-tethered sows. All behaviours which they considered to be abnormal were oral in nature. These totalled 11 (of 50) recorded actions (chewing/biting, sucking, mouth stretching, palate grinding, tongue flicking, licking, nibbling, nosing, rooting, pressing with rooting disc and pause) and were directed to 5 (of 9) possible substrates (trough, floor, bars, chain and nil). Up to 0.55 of stereotyped behaviours were performed with nil substrate, that is 'sham' behaviours. The mean number of fixed routines increased with stage of pregnancy until day 80 and then declined.

In a similar detailed study, Stolba et al. (1983) showed a strong increase in stereotyped behaviour over parities. Young sows showed high levels of 'drowsy' inactivity but also spent considerable time manipulating the limited amount of straw given. In second- and third-parity animals, time spent in straw-related behaviours decreased, but locomotory and investigative behaviours directed at the pen components increased and stereotyped behaviours started to develop. Older sows showed further increases in stereotyped behaviour, but all behavioural sequences became less variable with increasing age, and a higher proportion of behaviours were self-directed, that is 'sham' behaviours.

The development of stereotyped behaviour in individual tethered sows was traced by Cronin (1985). He showed that animals tethered for the first-time pass through a number of consistent behavioural stages. First comes a period of initial escape behaviour, in which the sow pulls on the tether, threshes about and screams. After 2-14 min, during which time escape attempts become shorter and less vigorous, this gives way to a period of inactivity. The animal lies immobile for long periods, making whining vocalisations. This phase typically lasts for about one day but with a range of 2 hours to 16 days. The animal then becomes more active again, displaying investigatory and aggressive behaviour. Initially a wide variety of such behaviours are performed, but over time there is a reduction in range of behaviour and basic stereotypies begin to develop over 8-55 days. Based on these observations Cronin (1985) suggested that the development of stereotypies was due to restraint and loss of control. However, an alternative hypothesis was developing at this time as a result of other studies.

4 Hunger in the pregnant sow

A number of studies had noted that abnormal oral behaviours occurred most frequently in the period around feeding time (Cronin, 1985; Blackshaw and McVeigh, 1984). Detailed studies carried out at this time showed that the type of behaviour varied in relation to the time of feeding. Head weaving, bar biting and snout rubbing were more commonly observed immediately prior to feeding, especially in older sows, whilst rooting, drinker manipulation and polydipsia were more common after feeding. Some sows spent as much as 40-60% of the first hour drinking. Other behaviour types (vacuum chewing, chain manipulation) were not associated with period of feeding. These observations led to the suggestion that stereotypies arose from the persistence of feeding motivation because commercially provided rations did not supply enough proprioceptive feedback, for example, stomach distension (Rushen, 1984, 1985). If stereotypies were due to frustration of feeding behaviour and not lack of environmental stimulation (as suggested for example by Stolba et al., 1983), the different forms of abnormal oral behaviour could be explained as motivationally appropriate, representing appetitive and consummatory phases of feeding. This suggestion was more in accordance with the generic models of behaviour being developed (e.g. Hughes and Duncan, 1988) which proposed that stereotypies result from the persistence of sequences of appetitive behaviour in restrictive environments which block negative feedback.

Further support for the pivotal role of feeding in causation of stereotyped behaviour came from studies of the consequences of altering the level or nature of feed. Gilts and sows tethered in unbedded pens spent more time standing and more time in repetitive behaviour when given low feed levels than those given high feed levels (Appleby and Lawrence, 1987). A subsequent study (Terlouw et al., 1991) showed that a similar effect could be shown in loose-housed animals (in an unbedded cubicle system). The incidence of chain manipulation and polydipsia was much more greatly influenced by feed level than by the degree of freedom of movement allowed by the housing system.

The role of gut fill in reducing stereotypies has been indicated in a number of studies. The provision of straw decreased the level of activity of tethered sows and the overall time spent in pen-directed oral activities (Fraser, 1975). In stall-housed sows, sham chewing, bar biting and apathetic behaviour were all reduced by provision of straw (Sambraus and Schunke, 1982). Providing additional roughage in the form of oat husks had a similar effect in stall housed sows (Broom and Potter, 1984). Even when less energy was consumed, feeding a diet which gave greater gastric distension reduced the occurrence of post-feeding oral behaviours (Brouns et al., 1994; van der Peet-Schering et al., 2003; de Leeuw et al., 2004, 2005).

Such findings led Lawrence and Terlouw (1993) to suggest that oral stereotypies result from the modification of foraging behaviour in highly food-motivated sows kept in behaviourally restrictive environments. Outdoor sows given lower feed levels increase the time spent foraging (Edwards et al., 1993a). In the absence of normal rooting substrates, such as soil, modified behaviours are performed. In many housing systems, straw provides an alternative substrate for rooting. Sows that were fed restricted diet spent more time rooting in their bedding than those fed on satiating high-fibre diets (Brouns et al., 1994). In a more critical analysis of the role of straw, Fraser (1975) showed that provision of long straw as bedding, allowing sows to manipulate it, reduced both the time spent standing and the proportion of standing time which was engaged in pen-directed oral activities, whereas providing the same quantity of straw chopped up in the food only reduced activity time and not the proportion of abnormal behaviours in this time. Providing long straw in the trough rather than as full bedding gave an intermediate result. This suggested that both nutritional and behavioural aspects of a fibrous substrate were important in alleviating the welfare challenges of restricted feeding. Subsequently, an interaction between feed level and availability of straw in the performance of abnormal oral behaviours was demonstrated. Sows fed on low levels and deprived of straw performed four times as much chain manipulation than those either given straw or fed at higher levels but without straw (Spoolder et al., 1995).

However, feeding is not the only activity influencing occurrence of abnormal behaviours in confined sows, since there are big individual differences between animals fed in the same way. This suggests that differences in individual coping style of the animals, as well as other external influences, are also important. Repetitive behaviour of newly tethered gilts was positively correlated with the level of chain manipulation performed by their neighbours (Appleby et al., 1989). This was most apparent in susceptible low-fed animals and may reflect an additive role of other environmental stressors such as disturbance from noise. The presence of an observer can also markedly increase the occurrence of abnormal behaviours (Sambraus and Schunke, 1982).

5 Pressure to adopt group housing systems for pregnant sows

Results such as those discussed previously have given rise to increasing public concern worldwide about the welfare of sows in confinement systems. In addition to the increased incidence of abnormal oral behaviours, other identified problems included a higher incidence of lameness and leg weakness (Svendsen and Bengtsson, 1983; de Koning, 1985), of urinary tract infections (Muirhead, 1983) and of prolonged farrowing and stillbirths (Nielsen et al., 1974; Svendsen and Bengtsson, 1983). Sows confined in pregnancy were also shown to have an impaired immune response (Metz and Oosterlee, 1980; Barnett et al., 1987a), and this may be one of the reasons for some reports of increased incidence of reproductive tract and mammary infections in the sow at farrowing, and of infectious disease in suckling piglets from sows housed in such systems (Backstrom, 1973; Svendsen and Bengtsson, 1983). However, such indications from these early studies (SVC, 1997) have not always been supported by more recent critical scientific review of the effects of system on sow health (Sow Housing Task Force, 2005).

As a consequence of public pressure, a number of countries have now enacted legislation which will restrict the housing of dry sows in individual confinement systems. Some countries, such as the United Kingdom, Switzerland, Sweden, Norway and Finland, have for many years had a total ban on individual confinement systems during gestation. The European Union banned tether housing from 2005 and from 2013 has restricted the use of gestation stalls to a period of four weeks after insemination. The New Zealand and Australian industries voluntarily announced similar partial stall bans in 2015 and 2017, respectively. In Canada any new pregnancy housing built from 2014 must allow for sows to be grouped, with a number of major retailers committing to stall-free supply chains by 2022. Several US states have also voted for a stall ban, and pressure from North American retailers to abolish this system is growing. Whilst such bans may solve one set of welfare problems for the pregnant sow, they have given rise to others associated with social grouping of sows.

6 Social organisation in sows

Sows are, in nature, socially living animals (Buchenauer, 1990). The wild pigs, from which modern domestic sows have evolved, live together in groups of 1–6 sows and their offspring. The group size depends on available resources such as food and sheltering cover. The typical social organisation for the species is therefore to have small groups of related animals, of mixed age and size, which have a relatively large home range and spend much of their time in foraging over this area. However, economic pressures dictate that commercial group housing systems operate with larger, often unstable, groups of unrelated, more uniform animals. These are subject to restricted space and concentrated feeding times and locations. It is the contrast of these two situations, where the biology of the animal can be at odds with the demands of the system, which gives rise to potential social problems.

Foremost amongst the problems of group housing is that of aggression between animals, which arises over competition for resources or when unfamiliar animals are grouped together. In stable groups, aggression is minimised by the use of subtle behavioural signals (Jensen, 1980b). In a detailed analysis of social interaction patterns in dry sows, he identified ten behaviour patterns. The function of these behaviours was investigated by an analysis of sequences of behaviours which occurred when two sows interacted (Jensen, 1982/3). The results indicated that head-to-head knock and head-to-body knock were aggressive attack behaviours, parallel and inverse parallel pressing were fight behaviours, nose-to-nose was a mild threat usually emitted by more dominant animals, nose-to-body and nose-to-genital appeared to be neutral behaviours associated with individual recognition and head tilt and retreat were submissive behaviours. Investigation of sows in semi-natural environments, where total space was greatly in excess of normal allowances, supported these results and identified a further pattern of behaviour called 'aiming', an upward thrust in the air with the snout, which appeared to be a mild threat behaviour (Jensen and Wood-Gush, 1984). Jensen (1982/3) proposed that in stable groups, aggression was regulated by an avoidance order. Lower ranking sows used head-tilt or retreat behaviours to avert attack in potentially aggressive situations. Comparison of sows in stalls, cubicles and pens with separate lying, dunging and feeding areas indicated that in the former two cases, where space was limited, these behaviours could not be adequately performed (Jensen, 1984). In consequence, confinement led to unsettled dominance relationships and increased aggression.

Deciding on what constitutes adequate space for group-housed sows is difficult since there are, at present, inadequate scientific data on this subject. It is relatively simple to calculate mathematical space requirements for individual animals in different postures from their physical dimensions (Petherick, 1983) and space envelopes when the animals are standing and lying (Baxter and Schwaller, 1983). Such information is useful in calculating dimensions for static situations such as resting or feeding in stalls, but this makes up only a part of the total space requirement. In addition to these static space requirements, the results previously discussed indicate that the animals require social space. Even in the absence of overt aggression, enforced proximity to another sow may in itself be stressful. Barnett et al. (1984) demonstrated a chronic stress response in gilts housed together in pairs, which was not seen in larger groups with greater total space. The preferred distance between individuals will vary, depending on the relationship of the animals involved and their current motivational state. Foraging sows in extensive conditions maintained an average distance of 3.8 m between group members and a distance of 50 m between different groups (Stolba and Wood-Gush, 1989). A distinction must always be made between space per animal and total space. With increasing group size, and constant space allowance per sow, the total pen area is increased, as is the potential for time sharing of space.

Investigation of the social structure of stable sow groups has shown this to be far from simple. In free ranging conditions, Jensen and Wood-Gush (1984) showed that the dominance order was more linear at feeding than during subsequent foraging. The number of interactions was also highest just after feeding. They suggested that a stable dominance order is not a prerequisite for low aggression level, but that this depends also on area and familiarity of the animals. In larger groups of mixed-parity sows, very linear hierarchies were found when all individuals were ranked relative to other group members by the results of paired food competition test (Brouns and Edwards, 1994). As noted in previous studies (Sambraus, 1981), rank was highly correlated to age/live weight of the animals, and the ranking obtained in pair testing was identical to that obtained by observations at feeding time within the group situation. Csermely and Wood-Gush (1986), observing group-fed sows in pens of 15, also found that the social hierarchy was the same during both feeding and non-feeding context. Most aggression was directed by high-ranking to low-ranking animals, with the next highest category being to sows of equal rank. Aggression from a subordinate to a dominant sow occurred only in non-feeding contexts. Investigation of groups of contemporary gilts, more even in age and weight, showed much less linear hierarchies (Stewart et al., 1993) in which rank was not correlated to live weight. Examination of the animals again in the second parity, when the groups had been reformed after being individually housed during lactation, showed that relative rank had changed very little. The linearity of hierarchy was influenced by the nature of the feeding regime. Animals fed individually in stalls, which were not therefore competing for food, showed a much less linear hierarchy than those which were group fed.

6.1 Establishing the dominance hierarchy

In a natural situation, where sows live in stable, mixed-parity family groups, the hierarchy can establish itself over time with a minimum of overt aggression. Young animals establish relative dominance with their contemporaries as early as formation of the teat order and, as immature growing animals, are naturally subdominant to older and larger sows within the family group. In contrast, in most commercial group housing systems, abrupt mixing of mature sows occurs at least once in each reproductive cycle. This occurs most often when groups are reformed after weaning. Sows may be regrouped with animals they were housed with a previous parity or with other animals not recently encountered. The extent to which sows recognise others from which they have been separated during a 4- to 5-week lactation period is uncertain, although Arey (1999) demonstrated that pregnant sows could be removed and returned into groups of six after a 6-week period without any major disruption to social organisation.

When unfamiliar sows are mixed, the number of interactions is high during the first 12 h, gradually decreasing over time (Edwards et al., 1993b; Csermely and Wood-Gush, 1990a,b; Luescher et al., 1990). After three days, the level of overt aggression is relatively low and focused around feeding times (Jones and Petchey, 1987; Csermely and Wood-Gush, 1986). Newly weaned sows, which show higher overall levels of activity, show more aggression than sows remixed in mid-pregnancy (Edwards et al., 1993b). Attempts to alleviate the aggression which occurs when animals are mixed have generally had little success, and there is still relatively little knowledge on the factors ameliorating aggression. Genetic differences in aggressive temperament of sows (Lovendahl et al., 2005), whilst undoubtedly important in this context, have not yet featured significantly in modern selection programmes. The effect of sedating the sows at mixing using pharmacological agents has been examined (Csermely and Wood-Gush, 1990a,b; Luescher et al., 1990), but, as with growing pigs, this seems only to postpone the onset of aggression and does not reduce the overall level of aggression. The use of odour-masking agents has also generally proved ineffective (Barnett et al., 1993b; Luescher et al., 1990). Provision of distractions such as fresh straw, whilst sometimes recommended commercially, has not been shown to be effective in a controlled experimental situation (Botermans, 1989).

The results from such studies indicate that some degree of fighting is inevitable when unfamiliar sows are penned together. However, the incidence or severity of fighting can be influenced by pen design. One option is to provide a large area for escape and avoidance of aggressors. Total space requirement in such situations is still poorly defined, and experiments with animals in unbedded pens have given conflicting results. When sows were mixed after weaning in groups of different sizes (3, 6 or 9 in a standard pen of 22.8 m²), sows in smaller groups, with more space per sow, avoided or withdrew from agonistic encounters more frequently and showed less severe aggression (Mujuni et al., 1985). However, in controlled experiments with gilts (Barnett et al., 1993a), reducing space allowance from 3.4 to 1.4 m² reduced the incidence of aggression in the first 90 min after mixing, although it had no effect on overall damage score after three days. Shape of the pen was also found to be important, since aggression was lower in a small rectangular pen and not in a small square pen. Similar results have been obtained with older sows in a comparison of groups of six mixed in pens providing 3.7 and 6.1 m² per sow (Edwards et al., 1993b). During the first 12 h there were more interactions in the large pen than in the smaller pen. These results appear to contradict the theory that a larger amount of space per sow reduces aggression. However, sows in the smaller pen in all cases exhibited more damage, suggesting that pen size may have more effect on severity of interaction as a result of limited escape possibility. In larger groups of 10, 30 or 80 sows in unbedded pens with floor feeding, Hemsworth et al. (2013) demonstrated a general decline in both aggression and plasma cortisol concentrations with increasing space over the range of 1.4 to 3 m², while there was a general increase in farrowing rate with increasing space. It is often suggested that sows in larger groups might require less space, since absolute pen size and therefore shared free space is greater. However, there were few interactions between group size and space allowance which would support this contention. In comparison of 2, 4 and 6 m² per sow, Greenwood et al. (2016) found no effect of space allowance on the number or duration of fights immediately after mixing, average lesion scores or subsequent reproductive performance. However, low-ranking sows had reduced lesions with greater space allowance, supporting the welfare benefit of a spacious mixing pen. Reducing space to 2 m² on day four, once aggression had subsided, had no adverse effects on any welfare or production measure.

The total space required to allow the interactions necessary to establish dominance with minimal injury is therefore uncertain, but it appears to be substantial. In one study where space was virtually unlimited, 0.75 of encounters resulted in chase distances of less than 2.5 m, but some sows were pursued up to 20 m following aggressive interactions (Edwards et al., 1986). Provision of such large areas would be uneconomic for mixing a small number of sows. However, it is possible that the provision of barriers which allow animals to visually separate themselves from an aggressor may partially compensate for reduction in total space. Constructing a pen with 'pop-holes' in which pigs could hide their head and neck during an aggressive interaction reduced aggression in newly weaned pigs (McGlone and Curtis, 1985). Attempts to mimic this effect in newly mixed gilts by providing partial stalls in the pen have proved unsuccessful (Luescher et al., 1990; Barnett et al., 1993a). However, the use of barriers has been successfully adopted in the design of an arena for initial mixing of unfamiliar sows (van Putten and van de Burgwal, 1990b). This Dutch design has been reported as allowing regular mixing of unfamiliar sows without serious injury, although comparative information on levels of aggression relative to alternative pen designs has not been published. In a UK study, provision of a central barrier in a pen reduced the number of interactions and was recommended, together with increased space and *ad libitum* feeding, as a component of a specialised mixing pen in which newly mixed groups could spend the first few days to minimise injury while social dominance was established (Edwards et al., 1994).

Once the dominance hierarchy has been established, the amount of space required for harmonious social interaction is still uncertain. This is probably because of the wide variation in pen designs, manure management, feeding systems and group sizes in use, with each of these factors affecting group dynamics, hygiene and the need for space. In small stable groups of six sows, with protected feeding in separate individual stalls, the quantity of skin lesions observed on the animals increased as space was reduced from 4.8 to 3.6, 2.4 and 2.0 m^2 per sow (Weng et al., 1998). A study of sows housed in groups of five with floor feeding, from day 25 of pregnancy during two consecutive parities, showed that animals had greater lesion scores and lower back fat thickness at 1.4 m^2 per sow than at greater specific allowances. However, there was no further improvement in these measures when space was increased from 2.3 to 3.32 m^2, although subsequent litter size was significantly higher at the greatest allowance (Salak-Johnson et al., 2007). In larger groups, more injuries were reported in dynamic groups in ESF housing at a space of 2.25 than 3.0 m^2 per sow, although no adverse effects on performance were seen (Remience et al., 2008). Thus, whilst good data are limited, the current EU legislation requiring a minimum of 2.4 m^2 per sow and 2.25 m^2 per gilt, with some adjustment for group size, seems sensible.

6.2 Dynamic grouping

Housing systems of lower capital cost, which are based on large group size, generally necessitate much more frequent mixing of sows, since the size of a contemporary weaning batch is only 4–5 animals for every 100 sows in the herd. Levels of aggression are higher in such dynamic groups than in smaller static groups which remain unchanged after their initial formation, although the degree of aggression recorded has varied widely in different situations. There can be significant performance loss if mixing stress occurs at critical times in the reproductive cycle (Lambert et al., 1986; Bokma, 1990; te Brake and Bressers, 1990). The extent to which the stressors associated with dynamic grouping impact on performance is still open to question, with different conclusions from different studies (Simmins, 1993; Anil et al., 2006) which probably reflect the subtle differences in management which can influence the outcome (Spoolder et al., 2009).

The factors affecting levels of aggression in dynamic groups are still poorly understood. It has often been suggested that aggression is lower when mixing occurs in larger groups. Whilst one experimental study has reported significantly less overall aggression in a larger group (32 v 21 sows), this was primarily related to accessibility of resources since there was little reduction in aggression towards newly introduced sows (Bokma and Kersjes, 1988). The key to successful functioning of dynamic groups appears to lie in the utilisation of subgroup behaviour. Within a large group of sows, the animals tend to form subgroups which lie together in specific areas of the pen (Edwards et al., 1986). If plentiful total space is available, and access to feeding and watering points well planned, newly introduced subgroups can remain together on the periphery of the main group with a minimum of aggressive interaction (Edwards et al., 1986; Hunter et al., 1989). Integration, a process shown to be effective when introducing gilts into groups of mature sows with little damage, can then take place gradually over a period of time (Lenskens, 1991). If total space is more restricted, as in partially slatted systems, there are benefits in creating specially partitioned lying areas for different subgroups to use (van Putten and van der Burgwal, 1990b). By closing off such an area for some days prior to introduction, the new group can move into an unclaimed resting territory and again avoid confrontation with longer established subgroups.

The degree of stress involved in mixing depends not only on the rank of an animal, but also on the strategy which it adopts in social encounters. Observation of the outcome of fights occurring when gilts were progressively added to a large group at 4- to 6-week intervals showed that it was possible to categorise animals into three groups (Mendl et al., 1992). High-success pigs were socially active, aggressive and won the majority of their encounters. No-success pigs never won encounters, were relatively inactive and showed low involvement in social interactions. Low-success pigs, which were of middle rank, were aggressive but relatively unsuccessful in social encounters. It was this third category which exhibited indications of physiological stress, with higher basal cortisol levels and greatest peak cortisol levels in response to ACTH challenge. This again highlights the possibility of selection for sow temperament in reducing the problems experienced in large dynamic group systems. In studies of stable groups of gilts, there was no correlation between response to ACTH challenge and rank (Edwards et al., 1993b). Thus, being of low rank may not in itself be stressful if a stable social situation exists and the frequency of aggressive interactions is low.

7 Aggression in stable groups and the method of feed provision

Once groups have been formed, the major factor giving rise to aggression in group-housed sows is competition for resources. It is normal commercial practice to feed pregnant sows a small amount of concentrate once or twice daily, and it is competition for this food which is the major cause of aggression in stable groups of sows (Carter and English, 1983; Jones and Petchey, 1987; Csermely and Wood-Gush, 1986). The one factor which has, above all others, given rise to problems in the design, cost and management of group housing systems for sows is the need to ensure that all animals obtain an adequate share of food without excessive aggression. Many different practical solutions to the problem of food distribution are in commercial use, with varying implications for sow welfare.

7.1 Floor feeding systems

The traditional, and cheapest, way of feeding group-housed sows is to distribute the total allowance of food on the ground and leave each individual sow to eat as much as it can until the food is finished. This may involve feeding by hand or be automated with dispensing canisters (Csermely and Wood-Gush, 1986, 1990a,b) (Fig. 2). However, this mode of delivery can give rise to aggression between sows as they compete for limited feed, both in an outdoor (Jensen and Wood-Gush, 1984) and an indoor situation, where up to 85% of all aggression-related behaviours can be food-related (Csermely and Wood-Gush, 1986; Jones and Petchey, 1987). The speed at which a concentrate diet is eaten varies widely between sows (Edwards et al., 1988a). Old sows can consume their 2–3 kg allowance in as little as 5 min, but younger sows may take up to twice as long. The inequality in food intake is further compounded by the ability of dominant sows to defend a particular area of good food supply (Csermely and Wood-Gush, 1990a,b), leaving lower-ranking individuals only limited access to the resource. Dominant sows occupied the centre of the feed pile, defending this area where food was thickest. They spent more time defending food than feeding, initiating more agonistic interactions than lower-ranking sows. Greatest aggression was seen in the first 15 min after food provision, with threat being more common in the subsequent 15 min. In consequence, these dominant sows showed a high frequency of short feeding bouts, whereas low-ranking sows had similar bout lengths but longer interruptions between bouts. These differences in feeding behaviour can have severe consequences for some animals. Brouns and Edwards (1994) showed that when sows were floor-fed, weight gain of low-ranking animals was only 0.63 of the group average in pregnancy. Large variation in body condition can result by the end of pregnancy when such systems are adopted in commercial practice

Figure 2 Group housing with floor feeding which gives aggressive competition for feed that can only be partly ameliorated by straw bedding.

(Edwards, 1992), and the increased aggression directed at low-ranking sows in such systems results in higher levels of skin lesions (Stewart et al., 1993).

7.2 Pens with individual feeding stalls

Precise rationing of each individual animal with minimal aggression can only be guaranteed by individually confining the animals at the time of feeding. The combination of group housing with individual feeding stalls offers a high welfare option (Edwards, 1985) (Fig. 3). Overall aggression is markedly reduced when feeding stalls are used in comparison with group feeding (Lambert et al., 1986; Jones and Petchey, 1987). In the latter study, it was noted that only 0.3 of the difference was due to aggression at feeding, indicating that the stalls may be beneficial in other ways. Although desirable for the sow, provision of individual feeding stalls which are used for only a short period each day may often not be adopted commercially because of high capital cost arising from stall purchase and additional housing space.

7.3 Cubicles and partial barriers

Space allowance and cost can be greatly reduced by combining the feeding stall and lying area, as is done in systems with cubicles or free access stalls (Edwards, 1985; Hoofs, 1990) (Fig. 4). However, as discussed earlier, such systems may result in undesirable effects on the pattern of social interactions if

Figure 3 Group pens with separate lying and activity areas and individual feeding stalls, which allow both socialisation and protected feeding but have higher requirement for space and capital cost.

group space is limited (Jensen, 1984). Further simplification of the pen can be made by reducing the length of feeding stalls, such that only partial barriers are used (Petherick et al., 1987). In a pilot study on such a system, where animals were fed a small amount of food twice daily, there was only a low incidence of poaching of feed and aggression at feeding was minimal. The feed bays were also used for resting and for retreat from aggression. However, there is little information about the functioning of such a system with larger groups of mixed parity, where it is likely that subdominant animals would suffer from food poaching and limited social space.

7.4 BioFix® or trickle-feed systems

One suggested way to minimise possible feed poaching by older, fast-eating sows, without the expense and inconvenience of closing all sows into full length stalls, is to use a 'biological fixation' (BioFix® or trickle-feed) system, in which feed is delivered by auger at the same controlled rate to each individual place. Since the sows cannot then eat at a differential speed, movement between feeding places gives no benefit and only short partitions along the trough are necessary to protect the feeding sows (Bengtsson et al., 1983; Hoofs, 1990; Hulbert and McGlone, 2006) (Fig. 5). Correct selection of the dispensing rate is essential to the success of the system, since too slow a rate will lead to restlessness in fast-eating sows, whilst too fast a rate will overwhelm the slow-eating animals. Experiments have shown that the number

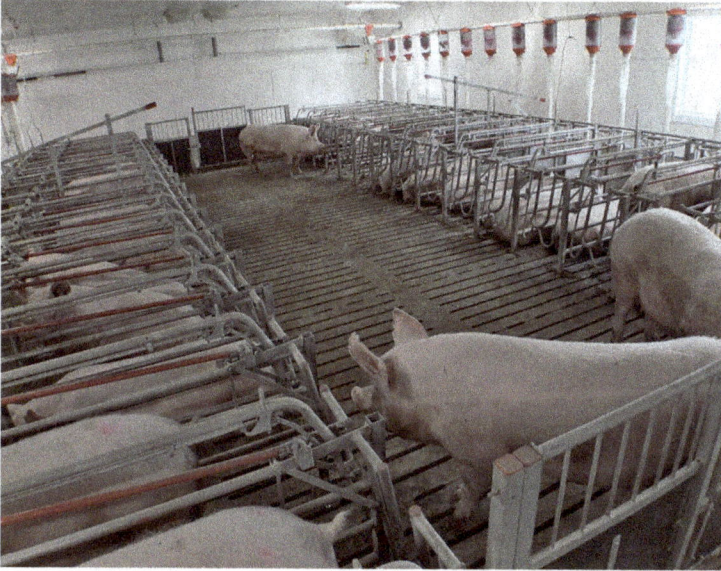

Figure 4 Free access stalls which reduce space requirement by combining the feeding and lying areas. Photo courtesy of Prairie Swine Centre, Canada.

of aggressive interactions and changes of place during feeding increase as the dispensing rate drops below 100 g of pellets per minute. However, with rates of more than 120 g per minute, more sows have food accumulating in the trough and the number of aggressive interactions when feed dispensing stops is increased (Hoofs, 1990). Sows fed at a slow rate by the BioFix® method subsequently spent less time nosing elements of the pen than sows given their feed in one total delivery (Bengtsson et al., 1983).

7.5 Two-yard systems

An alternative method for automated flat-rate individual feeding which can be used with simpler housing designs is the 'two yard' system (Hunter, 1988). Sows enter a mechanically operated feeding station sequentially from one yard and, after feeding, exit via a side gate into a different yard. Sows which have not fed remain in the first yard to await attention. Although simple in concept, some practical problems exist. Unless the size of the two pens is changed by moving of gates during the day, both pens must be large enough to house the majority of the group and total space requirement is high. The biggest problem is the possibility of mechanical malfunction and lack of precise information on the individual animal. Some sows pass through the feeder without stopping to eat (Hunter, 1988) and cannot then re-enter the feeder or be identified as not having fed.

Figure 5 Trickle-feed system which uses 'biological fixation' to reduce competition during feeding with only short stall protection.

7.6 Electronic sow feeding systems (ESFs)

It is well documented that the nutritional requirements of sows can vary widely depending on factors such as live weight, body condition and stage of pregnancy. No system of floor feeding or of flat rate feeding, for example, biological fixation or two-yard system, can meet the needs of every animal and some ability to differentially feed certain individuals is essential. In many practical situations, this is done by removing problem animals to an individual stall or pen for a period. However, automation of individual rationing was made possible in the early 1980s by the development of electronic sow feeders (Lambert et al., 1986; Edwards and Riley, 1986) (Fig. 6). In this system, animals are identified electronically by a device carried on a collar, ear tag or implant and must feed sequentially at one or more feeding stations controlled by a central computer. This enables large groups of animals to be kept in low cost, unspecialised housing (Brade et al., 1986). Sows in stable groups using this system soon develop a relatively stable feeding order, with dominant sows feeding at the start of the cycle and low-ranking sows waiting until a quieter time of day (Edwards et al., 1988b; Hunter et al., 1989). Sows newly introduced to the group generally begin low in the feeding order, progressively establishing themselves over time. They move up the feeding order as other longer-term group members are removed to farrow and new sows are introduced (Hunter et al., 1989). However, problems have occurred in practice with such systems. These have been mainly associated with design problems relating

Figure 6 Electronic feeding system where sows feed sequentially and can be individually rationed. Compartmentalisation of the lying area to encourage subgroup behaviour is also shown. Photo courtesy of Prairie Swine Centre, Canada.

to the feeding stall (Edwards et al., 1988b) and with the need for animals to feed sequentially. This is considered by some to be stressful for those animals frustrated in their attempts to feed and disruptive of the normal diurnal patterns of behaviour when animals feed throughout the night. It can also be a cause of aggression arising from competition for feeder entry. For this reason, the mechanical reliability of the gating system is of crucial importance (Edwards et al., 1988b), and the use of a walk-through stall design avoids the congestion at the feeder entrance which results with back-out stalls (Edwards et al., 1988b; Kroneman et al., 1993a,b). To reduce cost, simplified unprotected electronic sow feeding systems have been developed (e.g. FITMIX system) in which the complex stall and gating designs of conventional ESFs are absent. However, this results in a more competitive feeding environment to which some sows are unable to adapt (Chapinal et al., 2010).

Although simplistic conclusions about the relative welfare implications of these different housing systems are often made, it must be emphasised that the detail of system design and the management strategies adopted mean that sow welfare can vary as much between different versions of the same system as between the different systems (Edwards, 2000; Bench et al., 2013a,b).

7.7 Relationship between aggression and feed level

The behavioural problems encountered in group housing systems, like those of individual confinement housing, are frequently related to feeding practice. Increased levels of activity in low-fed animals may result in changes in social behaviour and aggression. It is a common observation that commercial units experiencing problems with high levels of aggression also have sows which are in poor body condition (Svendsen et al., 1990; Olsson et al., 1991). Providing food in a different form, or allowing the animals some other way to appropriately express their feelings of hunger, may alleviate such problems. Provision of a high-fibre diet reduced the number of skin lesions in floor-fed outdoor sows (Martin and Edwards, 1994). Spoolder et al. (1997) found no effect of feed level on aggression and skin lesions in an ESF system when the sows were bedded on straw, with sows on the lower feed level spending increased time in straw-directed foraging behaviour. Similarly, several studies of commercial group-housing systems have shown lower injury scores on farms with computerised feeding when generous occupational material (straw or hay) was provided (Weber et al., 1991; Gjein and Larssen, 1995). Vulva biting has been a major problem in many group-housing systems and particularly in ESF systems (van Putten and van de Burgwal, 1990a; Scott et al., 2009). There is good evidence that this abnormal social aggression may be related to feed provision. Dutch studies of sows in unbedded systems showed that vulva biting was most frequent after visits to the feed station when only small amounts (<1.0 kg) of concentrate were supplied, suggesting frustration at lack of satiation as the underlying cause. It has been demonstrated that a small quantity of food actually increases feeding motivation which, if the sow is left unsatiated, can be a major cause of reduced welfare (Terlouw et al., 1993). Supplying straw pellets in the feed station after a small amount of concentrate reduced the daily frequency of vulva biting and decreased the total number of wounded vulvas by 33% (Bure, 1991). Providing daily meals of chopped corn silage at the same time as the start of the feeding cycle also reduced the incidence of vulva biting from 30% to 10% (van Putten and van de Burgwal, 1990a). Where housing sows with straw bedding is not feasible, the presentation of straw or grass silage in racks may be beneficial (O'Connell et al., 2007; Stewart et al., 2008). However, such racks can become a focus for competition and aggression if poorly designed or sited (Kroneman et al., 1993a,b).

The problems associated with low feed levels and the cost of designing appropriate individual feeding systems indicate the desirability of providing a system in which the animal can feed *ad libitum*. Group-housed sows which are fed *ad libitum* do not eat at the same time, even when the possibility exists (Petherick et al., 1987) and feeding space can therefore be reduced in comparison to restricted systems. When fed *ad libitum* on conventional concentrate diets, dry sows will eat to excess and become obese, but this problem can be overcome by the use of specially formulated bulky diets of low nutrient density (Brouns et al., 1995; Ru and Bao, 2004) or low dry matter materials such as silages and by-products (Livingstone and Fowler, 1984; Edwards et al., 1992). Animals fed *ad libitum* in this way are less active, spend less time performing foraging behaviours indicative of hunger and have lower levels of skin lesions than animals competing for limited daily food (Brouns and Edwards, 1994; Stewart et al., 1993). Low-ranking animals are not disadvantaged by inability to obtain an adequate share of the food, as they can modify their feeding strategy to avoid conflict with more dominant animals by visiting the feeder at less popular times of a day. However, despite welfare advantages, this strategy is not widely adopted since the cost or handling logistics of these bulky diets can make it unattractive in many situations.

8 Extensive systems

The various problems described with intensive indoor housing systems for sows do not mean that welfare can only be good in an extensive situation, indeed the opposite can sometimes be the case (Edwards, 2005). In some countries, such as the United Kingdom, the use of outdoor systems for pregnant sows has increased in recent years, partly in response to market demand for perceived higher welfare product. Animal welfare is enhanced by the large area of free space, the environmental complexity and the choice of physical and social environment which is possible in these circumstances (Fig. 7). However, there is still competition for resources: notably for food, which is usually scattered once

Figure 7 Outdoor system for pregnant sows with simple hut shelter and feeding on the ground.

daily on the ground, and for shelter from extremes of both heat and cold at different times of a year. In these circumstances, the lower-ranking individuals within the group can experience severe welfare problems.

9 Conclusion

Many aspects of the housing and feeding of pregnant sows in commercial farms are at variance with the behaviour patterns which have evolved in response to the living conditions of their ancestors. This mismatch gives rise to a range of welfare problems for the animals. In individual confinement systems of stalls and tethers, the inability to resolve conflicts arising from enforced proximity to neighbouring sows and the inability to appropriately express foraging behaviour motivated by the chronic hunger associated with concentrated diets can result in chronic stress responses and the development of abnormal, stereotyped oral behaviours. These outcomes have given rise to public concern and led to the banning or restricted use of such systems in a growing number of countries. However, the adoption of group housing systems has brought a new set of welfare challenges linked to aggression between animals. Aggression occurs when unfamiliar animals are grouped together, when space is inadequate for the natural submissive behaviours to be appropriately expressed or when competition for resources occurs. Foremost amongst these is the competition for food in chronically hungry animals. Many different housing systems exist as a result of the multiple possible combinations of feeding system, floor type, bedding, group size and space allowance. Although generalised conclusions about the relative welfare implications of these different housing systems can be highlighted, it is actually the detail of system design and management on a case-by-case basis which has most influence on welfare outcomes. Reducing hunger without increasing calorie intake, by the use of high-fibre diets or supplementary forages, or allowing appropriate expression of foraging behaviour by the provision of straw bedding or other suitable substrate, gives welfare benefits in all housing systems. In group housing systems, the provision of adequate floor space for avoidance of aggressive encounters, the use of specially designed mixing pens with features assisting escape and hiding from more dominant sows and the facilitation of subgrouping behaviour in larger dynamic groups by compartmentalisation of the resting area can all minimise the social stress occurring in the system. Future strategies for high welfare in group housing must also take into account the genetic predisposition of the animals and exploit the potential for selective breeding for beneficial social genetic effects.

10 Future trends

Whilst much of the research to date has focused on housing design and management, future research will need to investigate more closely the sow factors associated with the successful implementation of group housing. The genetic and experiential effects on aggressiveness, whilst studied extensively in growing pigs, have as yet received insufficient attention in the sow. Lovendahl et al. (2005) calculated heritabilities of 0.17-0.24 for performance of aggression at mixing, suggesting significant scope for genetic approaches to improve welfare. The current work on selection for social genetic effects in growing pigs (Bijma, 2011) could be usefully extrapolated to gestating sows.

Another area where more research is required relates to the health challenges experienced by the pregnant sow. The issue of lameness has been highlighted in earlier studies (Kroneman et al., 1993a,b; Gjein and Larssen, 1995), but the change to group housing systems seems to have exacerbated the problems (Maes et al., 2016). Once again it is a multifactorial issue, with both animal and environmental (floor space, group size, pen design and flooring) contributing factors which need to be better understood in order to make appropriate recommendations. A new and concerning finding relates to the prevalence of gastric ulcers in pregnant sows. A survey by the Danish Pig Research Centre in 2014 showed that more than 50% of cull sows had damage to the gastric mucosa and more than 25% showed serious ulceration. The extent to which these ulcers cause pain to affected animals in relation to their degree of severity is unknown, although only animals with more severe ulcers show inappetence and loss of condition. In humans, the condition is known to be acutely painful, and the similarity in anatomy might suggest this to also be the case in pigs. If so, the high prevalence of the condition constitutes a serious welfare problem (Edwards, 2015). It is now widely recognised that the nature and quantity of the diet given to the pregnant sow are fundamental to many of the welfare issues in these animals, and further work on diets which can both promote satiety and support gut integrity is needed.

11 Where to look for further information

A description of different housing systems for pregnant sows can be found in the following papers:

Edwards, S. A. (1998). Housing the breeding sow. *In Practice*, 20, 339-43.

A series of 8 booklets on different group housing systems with discussion of their strengths and weaknesses and example housing layouts was produced by the UK Pig Welfare Advisory Group, and can be found at - http://webarchive.

nationalarchives.gov.uk/20031221024510/   and   http://www.defra.gov.uk/animalh/welfare/farmed/pigs/#pwag.

A useful website illustrating different systems and housing conversions is produced by the National Sow Housing Conversion Project (NSHCP) of Canada – http://www.groupsowhousing.com/.

The results of national reviews on the subject of welfare implications of housing for pregnant sows can be found in the following papers:

Scientific Veterinary Committee (1997). *The Welfare of Intensively Kept Pigs*. EU Commission.
EFSA. (2007). Scientific Opinion of the Panel on Animal Health and Welfare on a request from the Commission on Animal health and welfare aspects of different housing and husbandry systems for adult breeding boars, pregnant, farrowing sows and unweaned piglets. *The EFSA Journal*, 572, 1–13.
Sow Housing Task Force. (2005). A comprehensive review of housing for pregnant sows. *Journal of the American Veterinary Medical Association*, 227, 1580–90.
Bench, C. J., Rioja-Lang, F. C., Hayne, S. M., and Gonyou, H. W. (2013a). Group gestation housing with individual feeding – I. How feeding regime, resource allocation, and genetic factors affect sow welfare. *Livestock Science*, 152, 208–17.
Bench, C. J., Rioja-Lang, F. C., Hayne, S. M., and Gonyou, H. W. (2013b). Group gestation housing with individual feeding – II. How space allowance, group size and composition, and flooring affect sow welfare. *Livestock Science*, 152, 218–27.

Reviews of the issues associated with hunger in the pregnant sows can be found in the following papers:

Meunier –Salaun, M. C., Edwards, S. A. and Robert, S. (2001). Effect of dietary fibre on the behaviour and health of the restricted-fed sow. *Animal Feed Science and Technology*, 90, 53–69.
D'Eath, R. B., Tolkamp, B. J., Kyriazakis, I. and Lawrence, A. B. (2009). 'Freedom from hunger' and preventing obesity: The animal welfare implications of reducing food quantity or quality. *Animal Behaviour*, 77, 275–88.

Reviews of the issues associated with mixing of the pregnant sows can be found in the following papers:

Arey, D. S. and Edwards, S. A. (1998). Factors affecting aggression in newly regrouped sows and the consequences for reproduction. *Livestock Production Science*, 56, 61–70.

Greenwood, E. C., Plush, K. J., van Wettere, W. H.E. J. and Hughes, P. E. (2014). Hierarchy formation in newly mixed, group housed sows and management strategies aimed at reducing its impact. *Applied Animal Behaviour Science*, 160, 1–11.

12 References

Anil, L., Anil, S. S., Deen, J., Baidoo, S. K. and Walker, R. D. (2006). Effect of group size and structure on the welfare and performance of pregnant sows in pens with electronic sow feeders. *Canadian Journal of Veterinary Research*, 70, 128–36.

Appleby, M. C. and Lawrence, A. B. (1987). Food restriction as a cause of stereotyped behaviour in tethered gilts. *Animal Production*, 45, 103–10.

Appleby, M. C., Lawrence, A. B. and Illius, A. W. (1989). Influence of neighbours on stereotypic behaviour of tethered sows. *Applied Animal Behaviour Science*, 24, 137–46.

Arey, D. S. (1999). Time course for the formation and disruption of social organisation in group-housed sows. *Applied Animal Behaviour Science*, 62, 199–207.

Backstrom, L. (1973). Environment and animal health in piglet production. A field study of incidences and correlations. *Acta Veterinaria Scandinavica*, Suppl. 41, 1–240.

Barnett, J. L., Cronin, G. M., Winfield, C. G. and Dewar, A. M. (1984). The welfare of adult pigs: The effects of five housing treatments on behaviour, plasma corticosteroids and injuries. *Applied Animal Behaviour Science*, 12, 209–32.

Barnett, J. L., Winfield, C. G., Cronin, G. M., Hemsworth, P. H. and Dewar, A. M. (1985). The effect of individual and group housing on behavioural and physiological responses related to welfare of pregnant pigs. *Applied Animal Behaviour Science*, 14, 149–61.

Barnett, J. L., Hemsworth, P. H., Winfield, C. G. and Fahy, V. A. (1987a). The effects of pregnancy and parity number on behavioural and physiological responses related to the welfare status of individual and group-housed pigs. *Applied Animal Behaviour Science*, 17, 229–43.

Barnett, J. L., Hemsworth, P. H. and Winfield, C. G. (1987b). The effects of design of individual stalls on the social behaviour and physiological responses related to the welfare of pregnant pigs. *Applied Animal Behaviour Science*, 18, 133–42.

Barnett, J. L., Hemsworth, P. H., Newman, E. A., McCallum, T. H. and Winfield, C. G. (1989). The effect of design of tether and stall housing on some behavioral and physiological-responses related to the welfare of pregnant pigs. *Applied Animal Behaviour Science*, 24, 1–12.

Barnett, J. L., Hemsworth, P. H., Cronin, G. M., Newman, E. A. and McCallum, T. H. (1991). Effects of design of individual cage-stalls on the behavioral and physiological-responses related to the welfare of pregnant pigs. *Applied Animal Behaviour Science*, 32, 23–33.

Barnett, J. L., Cronin, G. M., McCallum, T. H. and Newman, E. A. (1993a). Effects of pen size/shape and design on aggression and injuries when grouping unfamiliar adult pigs. *Applied Animal Behaviour Science*, 36, 111–22.

Barnett, J. L., Cronin, G. M., McCallum, T. H. and Newman, E. A. (1993b). Effects of 'chemical intervention' techniques on aggression and injuries when grouping unfamiliar adult pigs. *Applied Animal Behaviour Science*, 36, 135–48.

Baxter, M. R. and Schwaller, C. E. (1983). Space requirements for sows in confinement. In: *Farm Animal Housing and Welfare* (Baxter, S. H., Baxter, M. R. and MacCormack, J. A. D. (Eds)). Martinus Nijhoff Publishers, The Hague, The Netherlands, pp. 181-95.

Bench, C. J., Rioja-Lang, F. C., Hayne, S. M. and Gonyou, H. W. (2013a). Group gestation housing with individual feeding - I. How feeding regime, resource allocation, and genetic factors affect sow welfare. *Livestock Science*, 152, 208-17.

Bench, C. J., Rioja-Lang, F. C., Hayne, S. M. and Gonyou, H. W. (2013b). Group gestation housing with individual feeding - II. How space allowance, group size and composition, and flooring affect sow welfare. *Livestock Science*, 152, 218-27.

Bengtsson, A.-C., Svendsen, J. and Persson, G. (1983). Comparison of four types of housing for sows in gestation: Behaviour studies and hygiene studies. Report 36, Swedish University of Agricultural Sciences, Lund, Sweden.

Bijma, P. (2011). Breeding for social interaction for animal welfare. In: *Encyclopedia of Sustainability Science and Technology* (Meyers, R. A. (Ed.)). Springer, Larkspur, CA, pp. 1477-1513.

Blackshaw, J. K. and McVeigh, J. F. (1984). The behaviour of sows and gilts, housed in stalls, tethers and groups. *Proceedings of the Australian Society of Animal Production*, 15, 85-92.

Bokma, S. (1990). Housing and management of dry sows in groups in practice: Partly slatted systems. In: *Proceedings of an International Symposium on Electronic Identification in Pig Production*. RASE, Stoneleigh, UK, pp. 37-45.

Bokma, S. and Kersjes, G. J. K. (1988). The introduction of pregnant sows in an established group. In: *Proceedings of the International Congress on Applied Ethology in Farm Animals*, Skara, Sweden, pp. 166-9.

Botemans, J. A. M. (1989). The effect of straw on the aggression of sows during grouping. Internal report, Swedish University of Agricultural Sciences, Lund, Sweden.

Brade, M. A., Edwards, S. A. and Riley, J. E. (1986). The commercial application of electronic sow identification and feeding systems in the UK. In: *Proceedings of the 37th Annual Meeting EAAP*, Budapest, Hungary.

Broom, D. M. and Potter, M. J. (1984). Factors affecting the occurrence of stereotypies in stall housed dry sows. In: *Proceedings of the International Congress on Applied Ethology*, Kiel, Germany.

Brouns, F. and Edwards, S. A. (1994). Social rank and feeding behaviour of group-housed sows fed competitively or ad libitum. *Applied Animal Behaviour Science*, 39, 225-35.

Brouns, F., Edwards, S. A. and English, P. R. (1994). Effect of dietary fibre and feeding system on activity and oral behaviour of group-housed gilts. *Applied Animal Behaviour Science*, 39, 215-23.

Brouns, F., Edwards, S. A. and English, P. R. (1995). Influence of fibrous feed ingredients on voluntary intake of dry sows *Animal Feed Science and Technology*, 54, 301-13.

Buchenauer, D. (1990). Social and feeding behaviour of sows in group housing systems. In: *Proceedings of the Seminar on 'Group housing of sows'*, European Conference Group on the Protection of Farm Animals, Brussels, Belgium, pp. 40-51.

Bure, R. G. (1991). The influence on vulva biting of supplying additional roughage in an electronic sow feeder. In: *Proceedings of the 42nd Annual Meeting EAAP*, Berlin, Germany.

Carter, A. J. and English, P. R. (1983). A comparison of the activity and behaviour of dry sows in different housing and penning systems. *Animal Production*, 36, 531-2.

Chapinal, N., Ruiz de la Torre, J. L., Cerisuelo, A., Gasa, J., Baucells, M. D. and Manteca, X. (2010). Aggressive Behavior in Two Different Group-Housing Systems for Pregnant Sows. *Journal of Applied Animal Welfare Science*, 13, 137-53.

Cronin, G. M. (1985). The development and significance of abnormal stereotyped behaviours in tethered sows. Doctoral thesis, University of Groningen, Groningen, The Netherlands, 146pp.

Cronin, G. M. and Wiepkema, P. R. (1984). The development and significance of abnormal stereotyped behaviours in tethered sows. *Annales de Recherche Veterinaire*, 15, 263-70.

Csermely, D. and Wood-Gush, D. G. M. (1986). Agonistic behaviour in grouped sows. I. The influence of feeding. *Biology of Behaviour*, 11, 244-52.

Csermely, D. and Wood-Gush, D. G. M. (1990a). Agonistic behaviour in grouped sows. II. How social rank affects feeding and drinking behaviour. *Bollettino di Zoologia*, 57, 55-8.

Csermely, D. and Wood-Gush, D. G. M. (1990b). Agonistic behaviour in grouped sows. III. The effect of grouping methods. *Bollettino di Zoologia*, 57, 271-5.

de Leeuw, J. A., Jongbloed, A. W. and Verstegen, M. W. A. (2004). Dietary fibre stabilizes blood glucose and insulin levels and reduces activity in sows (*Sus scrofa*). *Journal of Nutrition*, 134, 1481-6.

de Leeuw, J. A., Zonderland, J. J., Altena, H., Spoolder, H. A. M., Jongbloed, A. W. and Verstegen, M. W. A. (2005). Effects of levels and sources of dietary fermentable non-starch polysaccharides on blood glucose stability and behaviour of group-housed pregnant gilts. *Applied Animal Behaviour Science*, 94, 15-29.

De Koning, R. (1985). On the well-being of dry sows. Doctoral thesis, University of Utrecht, Utrecht, The Netherlands.

Edwards, S. A. (1985). Group housing systems for dry sows. *Farm Buildings Progress*, 80, 19-22.

Edwards, S. A. (1992). Scientific perspectives on loose housing systems for dry sows. *Pig Veterinary Journal*, 28, 40-51.

Edwards, S. A. (2000). Alternative housing for dry sows: System studies or component analysis? In: *Improving Health and Welfare in Animal Production*. EAAP Publication 102 (Blokhuis,, H. J., Ekkel,, E. D. and Wechsler,, B. (Eds)). Wageningen Pers, Wageningen, pp. 99-107.

Edwards, S. A. (2005). Product quality attributes associated with outdoor pig production. *Livestock Production Science*, 94, 5-14.

Edwards, S. A. (2015). Assessment and alleviation of pain in pig production. In: *Proceedings of the International Conference on Pig Welfare*. Danish Centre for Animal Welfare, Copenhagen, Denmark, 29-30 April.

Edwards, S. A. and Riley, J. E. (1986). The application of the electronic identification and computerised feed dispensing system in dry sow housing. *Pig News and Information*, 7, 295-8.

Edwards, S. A., Simmins, P. H., Walker, A. J. and Beckett, M. P. (1986). Behaviour of 400 sows in a single group with electronic individual feeding. In: *Proceedings of the International Symposium on Applied Ethology in Farm Animals*, Balatonfüred, Hungary.

Edwards, S. A., Armsby, A. W. and Large, J. W. (1988a). The effect of diet form on an individual feeding system with electronic identification for group housed sows. *Research and Development in Agriculture*, 5, 129-32.

Edwards, S. A., Armsby, A. W. and Large, J. W. (1988b). Effects of feed station design on the behaviour of group housed sows using and electronic individual feeding system. *Livestock Production Science*, 19, 511-22.

Edwards, S. A., Marconnet, C., Taylor, A. G. and Cadenhead, A. (1992). Voluntary intake and digestibility of distillery products for dry sows. *Animal Production*, 54, 486.

Edwards, S. A., Atkinson, K. A. and Lawrence, A. B. (1993a). The effect of food level and type on behaviour of outdoor sows. In: *Proceedings of the International Society for Applied Ethology*, Berlin, Germany.

Edwards, S. A., Mauchline, S. and Stewart, A. H. (1993b). Designing pens to minimise aggression when sows are mixed. *Farm Building Progress*, 113, 20-3.

Edwards, S. A., Mauchline, S., Marston, G. and Stewart, A. H. (1994). Agonistic behaviour amongst newly mixed sows and the effects of pen design and feeding method. *Applied Animal Behaviour Science*, 41, 272.

Gjein, H. and Larssen, R. B. (1995). Housing of pregnant sows in loose and confined systems–a field study. 2. Claw lesions: Morphology, prevalence, location and relation to age. *Acta Veterinaria Scandinavica*, 36, 433-42.

Fraser, D. (1975). The effect of straw on the behaviour of sows in tether stalls. *Animal Production*, 21, 59-68.

Greenwood, E. C., Plush, K. J., van Wettere W. H. E. J., and Hughes, P. E. (2016). Group and individual sow behaviour is altered in early gestation by space allowance in the days immediately following grouping. *Journal of Animal Science*, 94, 385-93.

Harrison, R. (1964). *Animal Machines*. Ballantine Books, New York, NY.

Hemsworth, P. H., Rice, M., Nash, J., Girri, K., Butler, L., Tilbrook, A. J. and Morrison, R. S. (2013). Effects of group size and floor space allowance on grouped sows: Aggression, stress, skin injuries and reproductive performance. *Journal of Animal Science*, 91, 4953-64.

Hoofs, A. (1990). Equipment assessment of group housing systems for sows in the Netherlands. In: *Proceedings of an International Symposium on Electronic Identification in Pig Production*. RASE, Stoneleigh, UK, pp. 77-82.

Hughes, B. O. and Duncan, I. J. H. (1988). The notion of ethological need, models of motivation and animal welfare. *Animal Behaviour*, 36, 1696-707.

Hulbert, L. E. and McGlone, J. J. (2006). Evaluation of drop versus trickle-feeding systems for crated or group-penned gestating sows. *Journal of Animal Science*, 84, 1004-14.

Hunter, E. J. (1988). Behaviour and welfare of dry sows in different housing conditions. Doctoral thesis, University of Reading, Reading, UK.

Hunter, E. J., Edwards, S. A. and Simmins, P. H. (1989). Social activity and feeder use of a dynamic group of 40 sows using a sow operated computerised feeder. *Animal Production*, 48, 643.

Jensen, P. (1980a). The impact of confinement on the behaviour of dry sows. An Ethological Study. Report 2, Department Animal Hygiene with Farrier's School, Skara, Sweden.

Jensen, P. (1980b). An ethogram of social interaction patterns in group-housed dry sows. *Applied Animal Ethology*, 6, 341-50.

Jensen, P. (1982/3). An analysis of agonistic interaction patterns in group-housed dry sows - aggression regulation through an 'avoidance order'. *Applied Animal Ethology*, 9, 47-61.

Jensen, P. (1984). Effects of confinement on social interaction patterns in dry sows. *Applied Animal Behaviour Science*, 12, 93-103.

Jensen, P. and Wood-Gush, D. G. M. (1984). Social interactions in a group of free-ranging sows. *Applied Animal Behaviour Science*, 12, 327-37.

Jones, R. H. R. and Petchey, A. M. (1987). The behaviour of dry sows in straw yards. In: *Pig Housing and the Environment* (Smith, A. T. and Lawrence, T. L. J. (Eds)). Occasional Publication no. 11, BSAP, Edinburgh, Scotland, pp. 134-6.

Kroneman, A., Vellenga, L., van der Wilt, F. J. and Vermeer, H. M. (1993a). Field research on veterinary problems in group-housed sows – A survey of lameness. *Zentralblatt für Veterinärmedizin. Reihe A*, 40, 704-12.

Kroneman, A., Vellenga, L., van der Wilt, F. J. and Vermeer, H. M. (1993b). Review of health problems in group-housed sows, with special emphasis on lameness. *The Veterinary Quarterly*, 15, 26-9.

Lambert, R. J., Ellis, M. and Rowlinson, P. (1986). An alternative sow housing/feeding system for dry sows based upon a sow-activated electronic feeder. In: *Proceedings of the 37th Annual Meeting EAAP*, Budapest, Hungary.

Lawrence, A. B. and Terlouw, E. M. C. (1993). A review of behavioral factors involved in the development and continued performance of stereotypic behaviors in pigs. *Journal of Animal Science*, 71, 2815-25.

Lenskens, P. (1991). The effect of the stepwise introduction of young gilts into an established group of sows on the aggression during grouping. Miscellaneous report 9, Swedish University of Agricultural Sciences, Lund, Sweden.

Livingstone, R. M. and Fowler, V. R. (1984). Pig feeding in the future: Back to nature? *Span*, 27, 108-10.

Lovendahl, P., Damgaard, L. H., Nielsen, B. L., Thodberg, K., Su, G. and Rydhmer, L. (2005). Aggressive behaviour of sows at mixing and maternal behaviour are heritable and genetically correlated traits. *Livestock Production Science*, 93, 73-85.

Luescher, U. A., Friendship, R. M. and McKeown, D. B. (1990). Evaluation of methods to reduce fighting among regrouped gilts. *Canadian Journal of Animal Science*, 70, 363-70.

Maes, D., Pluym, L. and Peltoniemi, O. (2016). Impact of group housing of pregnant sows on health. *Porcine Health Management*, 2, 17.

Martin, J. E. and Edwards, S. A. (1994). Feeding behaviour of outdoor sows: The effects of diet quantity and type. *Applied Animal Behaviour Science*, 41, 63-74.

McGlone, J. J. and Curtis, S. E. (1985). Behavior and performance of weanling pigs in pens equipped with hide areas. *Journal of Animal Science*, 60, 20-4.

Mendl, M., Zanella, A. J. and Broom, D. M. (1992). Physiological and reproductive correlates of behavioural strategies in female domestic pigs. *Animal Behaviour*, 44, 1107-21.

Metz, J. H. M. and Osterlee, C. C. (1980). Immunologische und ethologische Kritorien fur die artgemasse Haltung von Sauen und Ferkeln. *Kuratorium Tech Bauwesen Landwirtschaft Schrift*, 264, 39-50.

Muirhead, M. R. (1983). Pig housing and environment. *The Veterinary Record*, 113, 587-93.

Mujuni, B. M. K., McFarlane, J. M., Curtis, S. E. and Taylor, I. A. (1985). Effect of group size on social interactions in newly weaned sows. *Journal of Animal Science*, 63 suppl. 1, 164.

Nielsen, N. C., Christensen, K, Bille, N. and Larsen, J. L. (1974). Preweaning mortality in pigs. 1. Herd investigations. *Nordisk Veterinaermedicin*, 26, 137-50.

O'Connell, N. E. (2007). Influence of access to grass silage on the welfare of sows introduced to a large dynamic group. *Applied Animal Behaviour Science*, 107, 45-57.

Olsson, A. Ch., Andersson, M., Lenskens, P., Rantzer, D. and Svendsen, J. (1991). Herd studies of different electronic dry sow feeding systems. Report 75, Swedish University of Agricultural Sciences, Lund, Sweden.

Petherick, J. C. (1983). A biological basis for the design of space in livestock housing. In: *Farm Animal Housing and Welfare* (Baxter, S. H., Baxter, M. R. and McCormack, J. A. C. (Eds)). Martinus Nijhoff Publishers, Dordrecht, The Netherlands, pp. 103-20.

Petherick, J. C., Bodero, D. A. and Blackshaw, J. K. (1987). The use of partial barriers along the feed trough in a group housing system for non-lactating sows. *Farm buildings and Engineering*, 4, 32-6.

Remience, V., Wavreille, J., Canart, B., Meunier-Salaun, M., Prunier, A., Bartiaux-Thrill, N., Nicks, B. and Vandenheede, M. (2008). Effects of dry space allowance on the welfare of dry sows kept in dynamic groups and fed with an electronic sow feeder. *Applied Animal Behaviour Science*, 112, 284-96.

Ru, Y. J. and Bao, Y. M. (2004). Feeding dry sows ad libitum with high fibre diets. *Asian-Australasian Journal of Animal Sciences*, 17, 283-300.

Rushen, J. (1984). Stereotyped behaviour, adjunctive drinking and the feeding periods of tethered sows. *Animal Behaviour*, 32, 1059-67.

Rushen, J. (1985). Stereotypies, aggression and the feeding schedules of tethered sows. *Applied Animal Behaviour Science*, 14, 137-47.

Salak-Johnson, J. L., Niekamp, S. R., Rodriguez-Zas, S. L., Ellis, M. and Curtis, S. E. (2007). Space allowance for dry, pregnant sows in pens: Body condition, skin lesions, and performance. *Journal of Animal Science*, 85, 1758-69.

Sambraus, H. H. (1981). Das Sozialverhalten von Sauen bei Gruppenhaltung. *Zuchtungskunde*, 53, 147-57.

Sambraus, H. H. and Schunke, B. (1982). Verhaltensstorungen bei Zuchtsauen im Kastenstand. *Wiener Tierärztliche Monatsschrift*, 69, 200-8.

Scientific Veterinary Committee. (1997). *The Welfare of Intensively Kept Pigs*. EU Commission, Brussels, Belgium.

Scott, K., Binnendijk, G. P., Edwards, S. A., Guy, J. H., Kiezebrink, M. C. and Vermeer, H. M. (2009). Preliminary Evaluation of a prototype welfare monitoring system for sows and piglets (Welfare Quality project*). *Animal Welfare*, 18, 441-9.

Simmins, P. H. (1993). Reproductive performance of sows entering stable and dynamic groups after mating. *Animal Production*, 57, 293-8.

Sow Housing Task Force. (2005). A comprehensive review of housing for pregnant sows. *Journal of the American Veterinary Medical Association*, 227, 1580-90.

Spoolder, H. A. M., Burbidge, J. A., Edwards, S. A., Simmins, P. H. and Lawrence, A. B. (1995). Provision of straw as a foraging substrate reduces the development of excessive chain and bar manipulation in food restricted sows. *Applied Animal Behaviour Science*, 43, 249-62.

Spoolder, H. A. M., Burbidge, J. A., Edwards, S. A., Lawrence, A. B. and Simmins, P. H. (1997). Effects of food level on performance and behaviour of sows in a dynamic housing system with electronic feeding. *Animal Science*, 65, 473-82.

Spoolder, H. A. M., Geudeke, M. J., van der Peet-Schwering C. M. C., and Soede, N. M. (2009). Group housing of sows in early pregnancy: A review of success and risk factors. *Livestock Science*, 125, 1-14.

Stewart, A. H., Edwards, S. A., Brouns, F. and English, P. R. (1993). An assessment of the effect of feeding system on the production and social organisation of group housed gilts. *Animal Production*, 56, 422.

Stewart, C. L., O'Connell, N. E. and Boyle, L. (2008). Influence of access to straw provided in racks on the welfare of sows in large dynamic groups. *Applied Animal Behaviour Science*, 112, 235-47.

Stolba, A., Baker, N. and Wood-Gush, D. G. M. (1983). The characterization of stereotyped behaviour in stalled sows by information redundancy. *Behaviour*, 87, 157-82.

Stolba, A. and Wood-Gush, D. G. M. (1989). The behaviour of pigs in a semi-natural environment. *Animal Production*, 48, 419-25.

Svendsen, J. and Bengtsson, A.-Ch. (1983). Housing of sows in gestation. In: *Proceedings of the Guelph Pork Symposium*. University of Guelph, Guelph, Ontario.

Svendsen, J., Andersson, M., Olsson A. Ch., Rantzer, D. and Lundqvist, P. (1990). Group housing of sows in gestation in insulated and uninsulated buildings. Report 66, Swedish University of Agricultural Sciences, Lund, Sweden.

te Brake, J. H. A. and Bressers, H. P. M.. (1990). Applications in service management and oestrus detection. In: *Proceedings of an International Symposium on Electronic Identification in Pig Production*. RASE, Stoneleigh, UK, pp. 63-7.

Terlouw, E. M. C., Lawrence, A. B. and Illius, A. W. (1991). Influences of feeding level and physical restriction on the development of stereotypies in sows. *Animal Behaviour* 42, 981-91.

Terlouw, E. M. C., Wiersma, A., Lawrence, A. B. and Macleod, H. A. (1993). Ingestion of food facilitates the performance of stereotypies in sows. *Animal Behaviour*, 46, 939-50.

van der Peet-Schering C. M. C., Spoolder, H. A. M., Kemp, B., Binnendijk, G. P., den Hartog, L. A. and Verstegen, M. W. A. (2003). Development of stereotypic behaviour in sows fed a starch diet or non-starch polysaccharide diet during gestation and lactation over two parities. *Applied Animal Behaviour Science*, 83, 81-97.

van Putten, G. and van de Burgwal, J. A. (1990a). Vulva biting in group-housed sows: Preliminary report. *Applied Animal Behaviour Science*, 26, 181-6.

van Putten, G. and van de Burgwal, J. A. (1990b). Pig breeding in phases. In: *Electronic Identification in Pig Production. Proceedings of the International Symposium*. RASE, Stoneleigh, UK, pp. 115-20.

Vestergard, K and Hansen, L. L. (1984). Tethered versus loose sows: Ethological observations and measures of productivity. I. ethological observations during pregnancy and farrowing. *Annales de Recherches Veterinaire*, 15, 245-56.

Weber, R, Friedli, K. and Troxler, J. (1991). The influence of the computerised feeding system on the behaviour of sows and its effects with regard to injuries and alterations on their body. In: *Proceedings of the 42nd Annual Meeting EAAP*, Berlin, Germany.

Weng, R. C., Edwards, S. A. and English, P. R. (1998). Behaviour, social interactions and lesion scores of group-housed sows in relation to floor space allowance. *Applied Animal Behaviour Science*, 59, 307-16.

www.ingramcontent.com/pod-product-compliance
Lightning Source LLC
Chambersburg PA
CBHW050715280326
41926CB00088B/3038